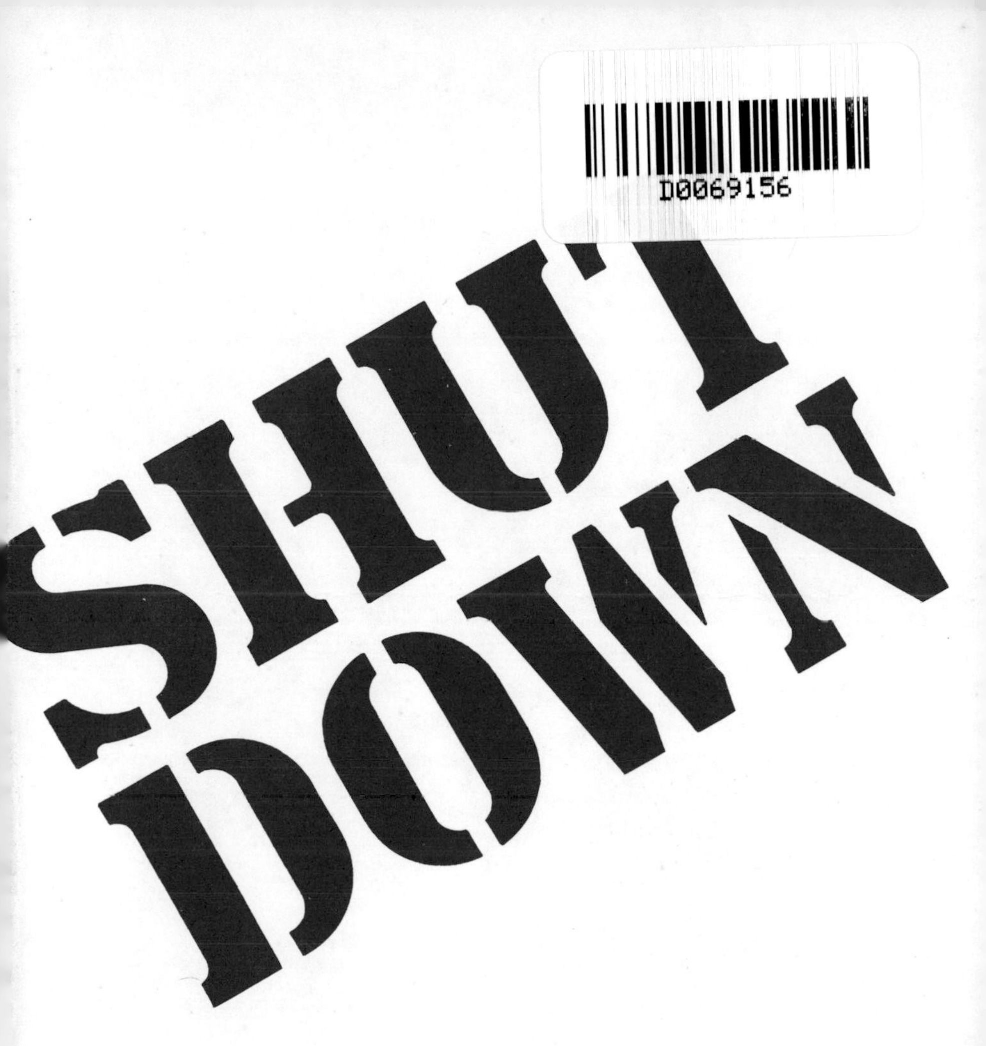

NUCLEAR POWER ON TRIAL

The Book Publishing Company – Summertown, Tennessee 38483

ISBN 0-913990-21-3

156 Drakes Lane, Summertown, TN 38483

TABLE OF CONTENTS

Preface

Two weeks after the first edition of ***SHUTDOWN***! rolled off the press, one of the most serious accidents in recent history occurred within a nuclear reactor on Three Mile Island near Middletown, Pennsylvania.

It had been a normal, quiet evening for the young, two-man crew that stayed in the control room through the Tuesday night "graveyard shift." Three Mile Island Unit 2 had been on line continuously for over a month. Two hours before the dawn of Wednesday, March 28th it was operating smoothly at 97 percent capacity. Then all hell broke loose.

Alarms sounded, warning lights began flashing, and Craig Faust, 32, and Ed Frederick, 29, were jolted into desperate action.

A water pump had failed and the lack of water had caused the turbine to stop turning. In just six seconds, pressure on the steam relief valve burst open with such an explosive force that it woke residents of the town of Goldsboro, across the Susquehanna River. The explosion blew radioactive steam high into the air with a long roar like a hot geyser.

Nine seconds into the accident the reactor tripped off automatically as pressures inside skyrocketed up, climbing 25 pounds per square inch every second. Steam temperature shot to the top of the scale.

In the first fifteen seconds Faust and Frederick pulled and pushed as many as fifty buttons and levers. But they needed more water to cool their rapidly overheating reactor; water that was unavailable because three hand-operated valves had been left closed by maintenance crews who serviced the reactor a few weeks earlier.

Faust and Frederick activated the Emergency Core Cooling System, but two minutes later it was shut off. The operators had made a crucial mistake. A pressure level indicator in the steam pipes leading away from the hot reactor had risen sharply and had gone off scale on the high side. The operators thought the system was flooded. Actually it was drying out. The pipes supplying water to the reactor core were shut. The steam relief valve which burst open to relieve the rising pressure now allowed radioactive water to overflow, flooding the floor of the reactor containment building.

The reactor core was in trouble. Its fuel rods were bowing and cracking from excessive heat. The coolant water had boiled off, leaving one third of the core uncovered. A red warning light on the control console was covered by a one-cent paper tag. Instrumentation in the control room suggested *too much* coolant water was passing through the core. There were *no* gauges which showed the actual level of water in the reactor.

With no water covering the fuel, the fuel tubes began to melt, spilling uranium and radioactive waste products into the reactor.

Finally, the Emergency Core Cooling System was reactivated and coolant water again flowed through the reactor vessel. As heavily contaminated boiling water spilled out into the containment building, radioactive steam surged into the atmosphere through open vents. Sump pumps automatically began piping the water on the floor to an adjacent structure known as the auxiliary building.

The auxiliary building was never designed to contain such heavy levels of radiation. While the reactor building, with three-foot thick concrete walls, was leaking radiation very slowly, radioactive water was being pumped rapidly from the auxiliary building into the Susquehanna River, and radioactive steam was blown out into the early morning sky. Once outside the buildings, the radioactive particles were picked up by variable winds and blown for miles. Authorities believed most headed north or northwest toward Harrisburg, Pennsylvania, but measurements showed increased levels of radiation in all directions out to twenty miles. A witness in Middletown said that the air had a metallic taste that morning.

When security guard Richard Bostdorf reported to work at the plant on Wednesday morning, no one told him anything was wrong. He punched in and proceeded toward the auxiliary building, where he worked. He heard a muffled noise from the containment building and the radiation alarm bell, and he realized something was wrong. At that point, he began to go through a standard operating procedure to secure the area. He came upon a supervisor who told him there was a problem, that they must evacuate and get the hell out of there. Everything was orderly but people were extremely concerned. In the process of securing the auxiliary building, Bostdorf waded through water up to his knees. No one told him that it was radioactive until he was monitored leaving the site. He was whisked into a mobile medical unit and showered immediately from head to toe in a chemical solution. His clothing was confiscated and destroyed. He was given a thin, paper outfit to wear.

He later told his son that he was amazed at the show of military force, the number of police, state police, and Army personnel who were there instantly. The company told the workers nothing, but it was obviously a very serious situation. The company told him not to talk to anyone about it.

Inside the control room the crisis was worsening. With a temperature of 150° at the bottom of the core and greater than 620° at the top, "you might infer that you have a situation where there is little or no flow in the core," said one of the men in the control room.

Three hours and fifteen minutes into the incident, instruments registered a small gas explosion in the reactor coolant drain tank, followed by a second explosion more than twice as large a few minutes later.

Pressure fluctuated wildly, dropping to 500 pounds per square inch, then shooting up to 2100 pounds per square inch. The instrument readings began to appear unreliable to the men trying to understand the problem in the core.

After nearly five hours of radioactive releases, the pumps leading to the auxiliary building were stopped and the reactor building vents were finally closed. The emergency response team from the Nuclear Regulatory Commission immediately detected high levels of radiation outside as they arrived.

Reactor operator Ron Fountain arrived in the control room on Wednesday to find 10 or 15 men wearing respirators on their faces. Some were strangers, talking on direct telephone lines to state and Federal officials. Fountain's immediate superior approached and asked if he would take a special assignment. He was to go into a radioactively "hot" area to turn one of the valves that was blocking cooling water.

Fountain suited up and entered the containment area. Although he was instructed to conserve oxygen in his respirator, as soon as he entered the area he found himself running to get done as fast as he could.

"I started hyperventilating," he said. "My instinct was to rip off my mask but I knew the area was heavy with particles. I said a prayer. I had to gather my wits. I was sweating and breathing heavy. I made my walk to the valve. I opened it. Then I walked toward the door." In about a minute, Fountain had received half his yearly maximum dose.

By midmorning on Wednesday the computer which normally prints out temperature readings in the core was printing out question marks, the sign that the computer has failed, or the thermocouple has failed, or the fuel rods are failing. The company technicians who had arrived at the scene said that the fuel rods were melting and that the temperature in the core had climbed so high that the computer could no longer read them. Hydrogen gas had escaped into the containment building and had exploded, with a pressure punch on the concrete walls of 28 pounds per square inch. The building's specifications were designed to withstand 40 to 45 pounds per square inch.

Outside the plant, the spokeman for the Metropolitan Edison company, Don Curry, was meeting with the press. There was "no nuclear accident" at the plant, he said, only a mechanical failure—a broken water pump. The Pennsylvania state official on the scene said, "Everything is under control. There is no danger to the public health and safety."

On Wednesday afternoon Metropolitan Edison called Edward Houser, the chemistry foreman at the plant, and asked him to report to work. When Houser reported, he was suited up with three sets of coveralls, three pairs of rubber gioves, a pair of rubber boots, and a full-face respirator. At 4:30 p.m. Houser walked into Unit No. 2 reactor containment building. He may have been the last person to walk into the building for some years to come. His assignment was to take a sample of the contaminated water on the floor.

Houser, who is married and has two children, stayed within the building for five minutes. He was later quoted as saying some of the readings were greater than 1,000 rem/hour. A 500 rem dose causes immediate death. After Houser left, twelve and one-half hours after the accident began, the containment building was sealed shut.

At 5:30 p.m., just before the evening news programs, the Nuclear Regulatory Commission issued a press statement which said, "The reactor is shut down. The pressure in the reactor system is being slowly reduced. NRC has a team of six persons at the site. They will participate in the NRC investigation of the event. The results will be made public." By Thursday morning radiation was still leaking from the plant, and the State of Pennsylvania began to send helicopter flights out to measure the radiation. The Environmental Protection Agency and the Food and Drug Administration also sent monitoring teams to Pennsylvania.

Thursday evening, with radiation still beaming through the concrete walls of the containment building and venting from the auxiliary building, an NRC spokesman on the site said, "At this time, the danger is over for the people off-site."

But all was not well within the control room. Nuclear Regulatory Commission supervisor on the scene, Roger Mattson, argued until midnight with the company operators, trying to persuade them that the control instrumentation revealed a serious situation. There was a growing bubble of non-compressible hydrogen in the core which was blocking the flow of coolant. Hydrogen is an explosive gas, and an explosion within the core could mean a loss of control over the reactor, resulting in a plant meltdown. The radiation of a thousand atomic bombs could be released by a large enough hydrogen explosion.

On Friday, Metropolitan Edison company Vice-president John Herbein told the press, "We don't believe an emergency exists to evacuate. . . I'm here today to try to lower the level of panic and concern."

The NRC's new representative on the scene, reactor systems analyst Harold Denton, was on the phone to Washington, talking to the Chairman of the NRC, Joseph Hendrie. "I think the important thing for evacuation to get ahead of the plume is to get a start rather than sitting here waiting to die," Denton told Hendrie. "Even if we can't minimize the individual dose, there might still be a chance to limit the population dose."

The public relations director for the NRC, Joe Fouchard, broke in: "Well, the governor is waiting on it, and Mr. Chairman, I think you should call Governor Thornburgh and tell him what we know... The Civil Defense people up there say that our state programs people have advised evacuation out to five miles in the direction of the plume. I believe that the Commission has to communicate with the governor and do it very promptly... Don't you think there should be some evacuation?"

Chairman Hendrie replied, "Probably, but I must say it is operating totally in the blind," Hendrie then called Thornburgh who wanted some authoritative word on the need for evacuation. Hendrie said it would be advisable to "suggest" that people stay indoors. Thornburgh was puzzled. He asked Hendrie, "Was your person, Mr. Collins, in your operations center, justified in ordering an evacuation at 9:15 a.m....? We really need to know that."

Hendrie replied, "I can't tell what the—I can go back and take a check, governor, but I can't tell you at the moment. I don't know."

Then Roger Mattson called in from the control center at the plant. He told Hendrie, "The latest burst didn't hurt many people. I'm not sure why you are not moving people. Got to say it. I have been saying it down here. I don't know what we are protecting at this point. I think we ought to be moving people."

White House Press Secretary Jody Powell called Fouchard in Pennsylvania and complained about the press coverage blowing the accident out of proportion. When Fouchard called Hendrie, the NRC Chairman said that he too had been speaking to the White House. Hendrie told Fouchard, "Now when I talked to Jody Powell a little bit ago we were concerned about having press conferences there at the site and then up here and people comparing tapes..."

Fouchard replied, "Harold [Denton] just talked very briefly with reporters here because there was no way we could hide him."

The Commission then came up with a press release and called the White House to read the final version to Powell. Back on the phone to Pennsylvania, Hendrie complained to operations director Lee Gossick about "having to deal with this media report that's going running from the UPI report and so on about meltdown being imminent and we are putting together, by the way, a press release that says no, there is not an imminent danger of a meltdown."

Gossick replied, "Yeah, I had a call from the White House situation room on that. I told them what had happened, that our guy had been taken out of context and misquoted." Hendrie acknowledged that was the correct procedure.

The situation at Three Mile Island was still basically out of control. Because a meltdown appeared possible, the Commissioners looked for some way that they could explain it. The situation was unlike any they had every foreseen. They yearned for something they were better able to handle.

"You know," suggested Hendrie, "what we need at the moment..."

"...is a good pipe break." completed Commissioner Peter Bradford.

Hendrie hoped aloud that one of the heavy electrical motors driving the reactor control rods "would just fall off the head of the damn vessel and give us a nice six-inch diameter small-break loss-of-coolant accident." The Commissioners then discussed Hendrie's idea, but some said they thought it would exacerbate the major problem of the moment, the hydrogen bubble in the reactor.

The "good pipe break" came up again on Saturday morning, when Mattson called to discuss ways to drive the hydrogen from the core. "We've got people looking at a way to fail a control rod drive on purpose and provide a crack. Unfortunately, the only way you can do that that we know of is to heat it; in other words you want to start a fire." Without deciding whether they wanted to start a fire, the Commission moved on to other subjects. A fire over a cracked control rod might make the hydrogen in the reactor core explode, bringing a meltdown.

Saturday morning a Metropolitan Edison official declared the crisis over and said the bubble had shrunk and that it contained only inert xenon gas. Harold Denton, the NRC analyst on the scene, was then interviewed separately and said the crisis was not over, the bubble was the same size, and that it consisted of explosive hydrogen gas. To avoid conflicting reports, a press blackout went into effect from officials at the Harrisburg site. In Washington the debate continued over how to respond to the accident.

Commissioner Richard Kennedy, who worked hard to tone down the press releases, said, "The focus, I think, has to be reassuring... reassure people that at least we are working on it." But after a staff technician explained how much pressure was building up in the vessel, Commissioner Victor Gilinsky said, "It sounds like the explosion is going to be a lot worse than we've let on."

Outside the government offices, nuclear critics were speaking up. MIT Professor Henry Kendall spoke for the Union of Concerned Scientists: "They are way out in an unknown land with a reactor whose instruments and controls were never designed to cope with this situation. They are like children playing in the woods." Ralph Nader, consumer advocate, called it "The beginning of the end of nuclear power." Dr. Ernest Sternglass arrived at the Harrisburg site and took his own radiation readings. He advised immediate evacuation of pregnant women and children under age five. Utility spokespeople attacked Sternglass as a crackpot whose estimates were worthless. Yet only a day after Sternglass' warning, Governor Thornburgh ordered an evacuation of pregnant women and pre-school children.

At nine o'clock Saturday evening, Harrisburg was shaken by reports that the gas bubble in the disabled reactor was on the verge of exploding. Ellen Hume, reporter for the *Los Angeles Times*, inquired, "Does this mean we are all going to die?" The report, it turned out, was based upon the casual remarks of an NRC official in Washington.

But sirens sounded in the streets. People ran to their cars and sped away. Roads became clogged with traffic as thousands fled in panic, fearing a catastrophic explosion. Extra police were brought into Middletown, a borough of 11,000, where the mayor said he would order his 13-man force to shoot looters. "I know that won't go over well with people, but that's the way I feel," said Mayor Robert Reed. "When you're fooling with the atom, you've got to take it seriously."

By midnight, Harold Denton and Governor Thornburgh had

gathered a press conference to announce, "There is no imminent catastrophical event foreseeable," and "It is certainly...many more days...before detonation limits would be reached." Later it was estimated that 50,000 to 250,000 persons had fled the 20-mile area.

Some refugees went to Hershey Park, Pennsylvania, where an evacuation center was set up in a sports arena built by the chocolate company. The refugees were upset and angry. "They ought to shut that damn thing down," said one man. A woman said, "You hear one thing from the utility, then one thing from the government, then another thing from Harrisburg and something else from civil defense. I don't know what to believe, what to do, so I guess the best thing is to go. It's better than doing nothing." She recalled arguing with her husband over living near a plant. "I just believed the company when they said it was safe. Now I don't believe it."

Marcella Baylor, age 8, was one of 83 children who slept through Saturday night on a cot on the ice hockey rink taken over by the Red Cross. "People are going to die from that stuff up in the sky," she told a reporter from Chicago. "My mom is scared to death."

On Sunday, President Jimmy Carter and an entourage of officials clad in thick yellow rubber boots toured the reactor site and reassured the public that the danger had passed. The workers in the control room, who had been wearing respirators to protect them from internal beta radiation hundreds of times more dangerous than the external gamma doses that were being reported, removed them for the President's televised visit, and donned them again after he left.

Unexpectedly Monday night, with the attention of much of the world focused on the tiny island in Pennsylvania, most of the hydrogen went back into solution in the cooling water to be extracted and vented out into the containment building. From containment the hydrogen was passed into recombiners and turned into water. On Tuesday, the reactor began to cool, slowly but steadily. The incident at Three Mile Island was over, turning the plant into what Senator Gary Hart called "a one billion dollar mausoleum." A spokesman for the Metropolitan Edison company said everything had gone exactly according to plan.

The accident in Pennsylvania is not the only one of its kind. Virtually every day, somewhere in the United States, a reactor containment building releases radiation to the environment. A core and containment cooling system fails about twice a day on the average in the 70 operating U.S. reactors. Not all accidents result in a loss of life, but almost all result in releases of radiation to the biosphere which, in turn, result in an accumulation of health effects.

On December 12, 1952, an explosion in a reactor at Chalk River, Canada, destroyed the core, leaking 4 million liters of radioactive coolant water.

A complete core was destroyed in a test reactor at Idaho Falls in 1954.

On November 29, 1955, the Experimental Breeder Reactor Number 1 in Idaho Falls, Idaho suffered a serious meltdown.

On September 11, 1957, a fire at the AEC Rocky Flats plutonium plant in Denver released plutonium to the Denver environs.

On October 8, 1957, Windscale-1 suffered a core burnup on the coast of England when 11 tons of uranium caught fire and 20,000 curies of iodine-131 escaped, contaminating 2 million liters of milk. Reactors 1 and 2 had to be filled with concrete. The milk was dumped into the Irish Sea.

In December, 1957, an accident involving waste storage at the Kasli Atomic Plant in Zyshtym, USSR, resulted in a "mud volcano" that cost thousands of lives and devastated hundreds of square miles of Ural Mountain farmland. Much of this area is still uninhabitable.

On May 25, 1958, a fire in the refueling room of the reactor at Chalk River, Canada, contaminated 400,000 square meters around the building.

Also in 1958, the core of yet another test reactor in Idaho Falls slumped from a partial meltdown.

In 1960 a single fuel element overheated and melted in Waltz Mill, Pennsylvania, forcing the test reactor there to shutdown.

On January 3, 1961, the SL-1, one of the seventeen reactors operated by the AEC in Idaho was shutdown for inspection and maintenance. On the night of January 3rd, three men were reassembling the control rod drives to prepare for restarting the reactor. Somehow the reactor went out of control. All three men were killed, one impaled by part of a control rod which pinned him to the ceiling of the containment building. The three men were so radioactive that they had to be buried in lead-lined caskets.

On June 11, 1962 an accident at the Uranium mill in Edgemont South Dakota spilled 200 tons of radioactive mill tailings into Cottonwood Creek. Much of the radioactive material washed 25 miles downstream until it sank into the Angostura Reservoir. No cleanup of the material has been undertaken. Between August of 1959 and September of 1977 there were sixteen of the same type of radioactive releases from U.S. uranium mills and none of these accidents have been cleaned up.

In April, 1963, the *U.S.S. Thresher* made a deep dive off Cape Cod and never resurfaced. The ocean was heavily contaminated with radioactivity. The death of the *Thresher's* entire crew was attributed to a reactor accident.

On October 5, 1966, a meltdown occurred in the Fermi Fast Breeder, imperiling Michigan. For many days atomic scientists worked to determine the extent of the damage and to bring the overheated core under control. Although radiation was escaping from the site, state and local officials were not notified of the danger. There were no existing evacuation plans. The reactor was finally brought to cold shutdown, the core was disassembled, and the building was placed under continuous security guard—mothballed for the present era. An engineer at the project said, "Let's face it, we almost lost Detroit."

In January, 1969, a loss of coolant accident occurred at Lucens, Switzerland, which seriously contaminated the reactor building, prohibiting access for two years. The reactor is now closed and the building used for waste storage.

In March, 1968, engineers cut a portion of pipe in the spent fuel storage coolant system of the Vallecitos reactor in California in order to relocate a pump. A deflated basketball was inserted into the pipe and inflated to close the line. When water pressure built up to 500 pounds, the basketball shot out of the pipe and 14,000 gallons of water drained from the spent fuel pool, pushing the level of radiation in the containment building up to 130 times higher than normal.

On May 11, 1969, one of the most expensive industrial fires in American history occurred at Rocky Flats plutonium facility in Colorado. 2,000 kg. of plutonium burned, much of it escaping in smoke to the surrounding hillsides. Of the over 200 fires in 20 years of Rocky Flats operation, this fire was the single most dangerous. In 1978, nine years later, the northwestern suburbs of Denver reported a doubling of the rate of lung cancer and birth defect mortality, and a tripling in the rate of leukemia. Sternglass has reported a four county cancer increase ranging from 44% to 92% for all cancers in the vicinity downwind of Rocky Flats.

On June 5, 1970 at Commonwealth Edison's Dresden II Nuclear Power Station in Morris, Illinois, the reactor went out of control for two hours after a meter gave a false signal and a monitor got its pen stuck. Radioactive iodine was released out of the containment building at one hundred times the permissible level.

On November 19, 1971, all of the waste storage space for the Northern States Power Company's Monticello, Minnesota reactor was filled and the company began spilling radioactive water into the Mississippi River. By November 21 about 50,000 gallons of wastes had been dumped into the river, and some were sucked into the domestic water intake for the city of Minneapolis before its gates could be closed.

On June 8, 1973, workers at the Hanford Atomic Works in Richland, Washington discovered a leak in one of the tanks storing high-level radioactive waste. The tank, built in 1944, had been draining by some 2500 gallons per day for 51 days. 115,000 gallons had escaped. This was only one of 18 accidents resulting in the loss of 549,400 gallons of high-level waste. There are more than 71 million gallons of this super-hot liquid waste still being stored "temporarily" at Hanford. While the 1973 leak did not contaminate groundwater, radioactive elements are slowly migrating toward the Columbia River.

On June 30, 1974, the AEC found for the previous year, a total of 3,333 safety violations at the 1,288 nuclear facilities it inspected. 98 posed a threat of radiation exposure to the public or to workers. The AEC imposed penalties for only 8 of these. In January, 1975, Congress abolished the AEC and replaced it with the NRC.

On March 22, 1975, at the Brown's Ferry Nuclear Plant near Decatur, Alabama, an electrician using a candle to search for air leaks in the cable-spreading room started a fire that burned for seven hours and destroyed 1,600 control cables, many connected to safety devices, including the Emergency Core Cooling System. Senate investigation revealed that the final reactor design had been approved even though it did not meet regulatory requirements. The plant came close to a meltdown. Total cost of the accident was placed at $150 million.

On June 6, 1975, at the Commonwealth Edison Zion reactor in Illinois, 15,000 gallons of radioactive water leaked into the containment building.

In July, 1976, a faulty valve at the Vermont Yankee station in Vernon, Vermont, caused 83,000 gallons of water contaminated with radioactive tritium to spill into the Connecticut River.

On December 14, 1977, after 14 separate explosions in the gas release system in GE plants around the U.S., Millstone I was rocked by two gas release system explosions, one of which blew an 80 pound steel door off the exhaust stack, released radioactive debris, and seriously exposed 50 employees, one critically.

On the night of January, 28, 1979, a truck carrying 46 barrels of radioactive waste from Illinois swerved to miss a slow-moving vehicle and overturned on a dark deserted mountain road in Tennessee. All the barrels spilled onto the highway and four ruptured. Civil defense workers who arrived at the scene took geiger-counter readings of 300 mR/hour, the maximum allowed by transportation regulations, but the truck was not marked with radiation placards because icy conditions had closed the highway for hazardous materials. The driver was drunk, and later jumped bail. An accident involving the transportation of radioactive materials happens an average of twice every week in America. One third of these accidents release radioactive materials to the environment.

In March, 1979, 21 Millstone employees were exposed to excessive radiation when 15,000 gallons of radioactive water overflowed a drain tank and spilled onto the floor of the building where the men were working.

In 33 years there have been over 10,000 disabling injuries and 300 deaths reported in the nuclear industry. On 23 occasions nuclear power systems have gone completely out of control. In the past seven years there have been over 150 threats against nuclear facilities, including three cases where bombs have been successfully planted. In 1979, a reactor destined for Iraq was destroyed at the factory in France. It was to have been a military reactor, fueled with weapons-grade uranium-235. After over three decades of nuclear energy experience, the amount of nuclear material inventory stolen or lost averages 100 pounds of uranium and 60 pounds of plutonium each year.

In an average year there are 3,000 reported accidents involving reactor safety systems in the United States. Equipment malfunction caused emergency shutdown of reactors 127 times in 1977. 12% of all reported accidents involve leaks or airborne releases of radiation.

The remarkable thing about the nuclear industry is that accidents need never occur for citizens to lose their lives and for the environment to become contaminated with radiation. Every day, the nuclear fuel cycle releases routine emissions of radioactive gases and liquids and has small accidents which spill long-lived radionuclides into the environment. These toxic elements will result in an increased rate of cancer and birth defects, as well as a broad spectrum of diseases.

Decisions which must be made as to the potential uses of nuclear power in the coming years should be made with a full understanding of its effect on the public health. The radiation released at Three

Mile Island is the same kind of radiation that is released by an atomic bomb. Radiation was distributed in the same kind of fallout pattern over Pennsylvania as the kind found after nuclear weapons tests in Nevada. The routine releases of low-level radiation at nuclear plants are currently causing diseases and genetic effects whose full impact will not be felt for centuries.

Our society is engaged in a critical re-evaluation of nuclear power as an energy source, and the testimony in this book, by two of the leading spokesmen for the scientific community, will prove valuable for an understanding of the health effects of low level radiation and the results which we may expect from our commitment to develop nuclear power.

Clifford Alle

Introduction

The late Congressman Clifford Allen of Tennessee spent his last Thanksgiving Day composing a press release about the severe underestimates of radiation released into the biosphere from the nuclear fuel cycle. He had just received some alarming information, a copy of the memo written by Dr. Walter Jordan, a member of the Atomic Safety and Licensing Board and a former Assistant Director of Oak Ridge National Laboratories. In 1977, in what has come to be known as the "Jordan Memorandum," Dr. Jordan disclosed that the estimates of the releases of radon gas from the nuclear fuel cycle had been 100,000 times too low. Dr. Jordan's figures showed that as many as one hundred deaths could eventually result *from each day* that the nuclear power industry continued in operation.

With Congressman Allen as he composed his press release was Jeannine Honicker, a Nashville businesswoman. Jeannine's daughter, Linda, had contracted leukemia at age nineteen, but recovered after a difficult and complicated bone marrow transplant. Jeannine's husband, Dolph, News Editor for the Nashville *Tennessean,* had written Linda's story for the *Reader's Digest.*

In the process of learning about leukemia, Jeannine discovered something else. Leukemia is one disease which has been shown to be caused by radiation. According to health physicists, a doubling of the spontaneous rate of leukemias might be part of the price we would pay if we used nuclear-generated electricity. Jeannine was among more than thirty intervenors in the licensing process for the world's largest nuclear plant at Hartsville, Tennessee. Joining with nuclear opponents in ten southern states, she helped to found Catfish Alliance. Following Clifford Allen's death, she ran for his seat in Congress, unsuccessfully.

In early 1977 Jeannine met Stephen Gaskin, founder of The Farm, a religious community in Summertown, Tennessee, and Albert Bates, a paralegal associated with Farm Legal. They agreed to help prepare a case against the Nuclear Regulatory Commission.

On July 29, 1978, a Petition for Emergency and Remedial Action was filed with the Nuclear Regulatory Commission, citing

Jeannine Honicker

the Jordan Memorandum and other recent government disclosures and asking that the Commission suspend the licenses of the nuclear industry pending a complete investigation of the biological effects of low-level ionizing radiation.

When the Commission did not respond on an emergency basis, a suit was filed in Federal Court in Nashville, seeking an injunction to shut down the nuclear fuel cycle. A telegram was sent to the Nuclear Regulatory Commission, notifying them that Jeannine would seek an order restraining them from violating her civil rights and damaging the public health.

A temporary restraining order was requested on the basis of the NRC's own figures concerning releases of radiation and the likelihood that a large number of people might eventually die from each day of continued operation. The Court declined to issue the temporary restraining order but set a preliminary evidentiary hearing for October 2, 1978.

The complaint filed in Federal Court asked the Judge to order the members of the U.S. Nuclear Regulatory Commission to revoke the licenses of all nuclear fuel cycle facilities.

Two of the most respected authorities on radiation sciences appeared at this October 2nd hearing as witnesses for Jeannine Honicker. They were Dr. John W. Gofman, Professor Emeritus of Biology at the University of California, Berkeley, and Dr.

Ernest J. Sternglass, Professor of Radiation Physics at the University of Pittsburgh. Their testimony lasted about four hours. A third witness, Dr. Chauncey Kepford, traveled from Pennsylvania with Dr. Sternglass but was not allowed to testify on this occasion. Dr. Kepford would have testified as to the Jordan Memorandum and the underestimated effects of the radon gas emitted wherever uranium is processed.

The first decision in this case went back to 5,000-year-old Biblical Law. October 2, 1978, was also *Rosh Hashanah*, the Jewish New Year. Attorneys for the Nuclear Regulatory Commission had asked the Court to postpone proceedings, since two of their attorneys and at least one material witness were of Jewish faith. Dr. Sternglass and Jeannine's attorney, Joel Kachinsky (both also Jewish), were opposed to the delay: A life and death matter should take precedence over even a High Holiday.

Judge Morton dismissed the NRC motion and held proceedings as scheduled on October 2. Except for some unnecessary repetitions and digressions, the following is a record of those proceedings.

You are invited to judge for yourself.

Judge: Honorable L. Clure Morton, Chief Judge
Nashville United States District Court

For the Plaintiff:

Plaintiff: Jeannine Honicker
Farm Legal: Joel Kachinsky, Albert Bates

For the Defendants:

Nuclear Regulatory Commission: Stephen Eilperin, Leo Slaggie, Sheldon Trubatch
United States of America: Irvin Kilcrease, Jr.

TESTIMONY

TRANSCRIPT OF PROCEEDINGS

October 2, 1978

BY THE CLERK: Case No. 78-3371, NA-CV, Jeannine Honicker versus Joseph M. Hendrie, et al.

Is the plaintiff ready?

BY MR. KACHINSKY: The plaintiff is ready.

BY THE CLERK: Is the defendant ready?

BY MR. KILCREASE: Defendant is ready. If Your Honor please, may I approach the podium?

BY THE COURT: All right.

BY MR. KILCREASE: For the record, I am Irvin H. Kilcrease, Jr., Assistant United States Attorney, and, if Your Honor please, at this time I would like to make some introductions of attorneys who will be, with the Court's permission, will be representing the Nuclear Regulatory Commission.

If they will come around at this time. If Your Honor please, I move the admission of the following attorneys for the specific purpose of representing the Nuclear Regulatory Commission and the United States government in this case.

BY THE COURT: You mean they don't trust you?

BY MR. KILCREASE: I'm just going to be sitting with them.

The first one on my left, if Your Honor please, is Steven F. Eilperin. He's a member of the Bar of the State of New York.

BY THE COURT: Will you spell his name please?

BY MR. KILCREASE: E-i-l-p-e-r-i-n, Steven F. of the Bar of New York and Federal District Court of the District of Columbia.

The next person is Mr. Leo Slaggie. He's a member of the Supreme Court of the State of California.

And the third person is Mr. Sheldon L. Trubatch. He's a member of the Supreme Court of New York.

BY THE COURT: Motion granted. Delighted to have you, gentlemen.

BY MR. KILCREASE: If your Honor please, one other matter, Mr. Leo Slaggie has a hearing problem, and we would ask permission of the Court that he be permitted to move around to positions where he can hear various witnesses.

BY THE COURT: He can sit—he can designate one place where he's going to sit and stay there.

BY MR. KILCREASE: All right.

BY THE COURT: If he wants to sit over here, we will draw him a chair up over here, but now

I'm not going to have people running around the courtroom back and forth. It interferes with my sleep.

BY MR. KILCREASE: All right, I will confer with him and find out where the best position is.

BY THE COURT: I would suggest—we will turn the hearing thing up substantially, but I would suggest probably right over here in a seat right here would be where the Marshal usually sits. The Marshal will bring up another chair over there, and he can keep the Marshal awake.

All right, anything further?

BY MR. KILCREASE: Yes, sir, we filed a motion to dismiss on September 29th, this past Friday.

There was a memorandum in support of that motion, and we stated in the motion that the exhibits mentioned in the memorandum would be produced at this hearing.

I furnished Plaintiff's counsel with a copy, and I want to pass this—file this with the Clerk.

BY THE COURT: All right, sir.

BY MR. KILCREASE: All right, I'm asking the Marshal to pass—this is not an exhibit. It is just a pertinent part of the Federal Register, Volume 43. I passed it around. I want to present the Court with a copy—

BY THE COURT: All right.

BY MR. KILCREASE: —for the Court's convenience.

I have conferred with Mr. Slaggie. He would like to stay where he is until the plaintiff puts their first witness on.

BY THE COURT: That's right now. We're not going to have opening statements. They are going to put any proof they have on right now.

BY THE CLERK: For the record, Exhibit 1 is a letter to Mrs. Jeannine Honicker.

BY THE COURT: And various attachments.

[*Marked and filed Exhibit Nos. 1-4 in evidence*]

BY THE COURT: Call your first witness, Put on your proof.

BY MR. KACHINSKY: Do I need to make an opening statement?

BY THE COURT: No, just put on your proof.

* * * * *

DR. JOHN WILLIAM GOFMAN, having been first duly sworn, was thereupon called as a witness and testified as follows, to wit:

DIRECT EXAMINATION

BY MR. KACHINSKY: I have fourteen exhibits I would like to have the Clerk mark that I will use in questioning the witness.

BY THE COURT: Have you shown those exhibits to defense counsel?

BY MR. KACHINSKY: No, Your Honor.

BY THE COURT: Pass them over. Let defense counsel look at them right now.

BY MR. KACHINSKY:

Q. ***Would you please tell the Court your name and address?***

A. My name is John William Gofman, G-o-f-m-a-n, and my address is 1045 Clayton Street, San Francisco, California.

Q. ***And what is your occupation?***

BY THE COURT: Mr. Kachinsky, have you read the rules, the local rules of this Court that are in printed form?

BY MR. KACHINSKY: Yes, I have, Your Honor.

BY THE COURT: Those rules provide that you will give a summary of all of this background information and ask one question and ask the witness if that's correct. Do you have all that information that you can recite?

BY MR. KACHINSKY: Yes, I do, and Exhibit No. 5 is a biographical sketch.

BY THE COURT: Let Exhibit No. 5 be filed as an exhibit and go on to a pertinent question then. Let Exhibit No. 5—do you have it over there? Let it be filed. I will read that. You don't have to go into that. Exhibit No. 5 is a vitae, *all right.*

[***Marked and filed Exhibit No. 5 in evidence***]

BY THE COURT: Go ahead, sir.

BY MR. KACHINSKY: [*Continuing*]

Q. ***Okay, as a result of your education and experience in relevant areas, do you feel qualified to answer questions on radiation physics and biology?***

A. Yes, Sir.

Q. ***Could you tell us briefly what is radiation?***

Biography of John W. Gofman, M.D., Ph.D.

Education

A.B. Chemistry, Oberlin College, 1939
Ph.D. Nuclear Chemistry, Univ. of Calif., Berkeley, 1943
M.D., School of Medicine, Univ. of Calif., San Francisco, 1946
Internship in Internal Med., Univ. of Calif. Hosp., San Francisco, 1946-47

Positions

Academic appointment, Div. of Medical Physics, Dept. of Physics, U.C. Berkeley, 1947; advancement to Full Professor, 1954; Emeritus, 1973.
Concurrent appointment (1947 on), Instructor or Lecturer in Medicine, Dept. of Medicine, Univ. of Calif., San Francisco.
Medical Director, Lawrence Livermore Laboratory, 1954-1957; Associate Director, 1963-1969.
Founder and First Director, Biomedical Research Division, Lawrence Livermore Laboratory, 1963-1965.
Chairman, Committee for Nuclear Responsibility (current).

Honors and Awards

Gold-Headed Cane Award, 1946, to graduating senior for qualities as a physician, U.C. Medical School
Modern Medicine Award, 1954, for outstanding contributions to heart disease research
Lyman Duff Lectureship Award (Amer. Heart Assoc.), for research in atherosclerosis and coronary heart disease
Stouffer Prize, 1972, for outstanding research contributions in arteriosclerosis
One of the 25 Leading Cardiologist Researchers of the Past Quarter-Century, American College of Cardiology, 1974

Patents

- Discovery of Fissionability of Uranium-233
- Two Processes for Isolation of Plutonium

Books Published

What We Do Know About Heart Attacks
Dietary Prevention and Treatment of Heart Disease (with A.V. Nichols and V. Dobbin)
Coronary Heart Disease
Population Control through Nuclear Pollution (with A. Tamplin)
Poisoned Power: The Case Against Nuclear Power (with A. Tamplin)

Other Publications

Approximately 150 scientific articles encompassing the following fields:

- Lipoproteins, atherosclerosis, coronary heart disease
- Trace elements by X-ray spectroscopy
- Chromosomes and cancer
- Medical effects of ionizing radiation
- Nuclear power, the hazards of plutonium and other sources of ionizing radiation

A. Radiation is one form of energy. We have radiation ranging in wave length all the way from very long wave length to very short wave length, and the types of radiation we are concerned about here are those of very short wave length in the form of X-rays and gamma rays; and in addition these can be generated by machines, for example, X-ray generators, or they can come from natural and man-made substances.

In addition to that form of radiation, we have particles that can be emitted by radioactive substances, such as electrons, which we call either beta rays or positrons.

We have alpha particles which are charged nuclei of helium.

These are all forms of radiation; either waves in nature or particles. Actually the waves are also regarded as particulate for some purposes.

Q. ***Okay. How does radiation affect living organisms?***

A. In general, ionizing radiation affects living organisms in a destructive manner. It causes, as it goes through the cells of living organisms, the ripping away of electrons from the molecules or atoms in which they are present and thus altering those atoms and molecules to some other form.

In addition to ripping away electrons from atoms and molecules, it can often displace electrons from one energy state in the molecule to another. All of these have the effect of altering the naturally occurring substances in a biological organism.

Q. ***And what is the effect of radiation on cells and genes?***

A. There are probably many effects. The major one that we are concerned about in cells is upon the genetic material or genes which reside in the nucleus of cells in the form of a long chain-like molecule that is twisted around and is called DNA, and radiation can break that chain, which is one very important effect. It can also alter the chemical structure of some of the submolecules of that chain, and if those submolecules of the chain are altered, the information contained in the cell's genetic environment or genetic endowment is simply changed and it won't do the right things thereafter. It would do different things from what it normally does.

Q. ***And how is radiation connected to cancer?***

A. We cannot be sure of the mechanism by which radiation is connected to cancer, but we can be absolutely certain from the evidence that has been adduced that radiation is one of the causes of human and animal cancer.

As I said, we do not know the exact mechanism. The leading speculations are that radiation by its ability to either break or rejoin in an abnormal way these chains of genetic information in the cell lead to an alteration in the control mechanism for the cells.

Ordinarily human organisms and other animals are very remarkable in that we do not have cells going on to reproduce wildly.

A man's liver grows to a certain size, and, indeed, when some cells are injured, it replaces those cells, but it doesn't replace an infinite amount of those cells. The same is true for the lining of the intestinal tract. The same is true in the bone marrow for the blood cells.

We think of cancer primarily as a cell that no longer responds to the control mechanisms that tell it not to keep reproducing, and then we get a mass growing and invading other tissues.

We think the information in the cell to cause this non-proliferation when it's not needed to be a control mechanism that is in those chains that I spoke of, and the genes are organized into forty-six structures in a normal cell in humans, called chromosomes.

It is entirely possible, though not proven, that one of those chains is the regulator of telling the cell when to divide and when not to; and if you injure that regulator, that cell no longer has the information to tell it not to divide in appropriate circumstances, and that itself can be cancer or leukemia.

However, I would like to say it is a speculation that this regulator gene mechanism is *the* mechanism; but most scientists think the defect that leads to cancer is some injury to the genes or chromosomes in the nucleus of the cell.

Q. ***And how long would it take for this cancer to develop?***

A. If the injury to a cell, for example, when you were irradiated is immediate within fractions of a second, that injury is there, and the type of injury that can lead to cancer has been produced and is essentially irreversible. Certain forms of injury are reversible, but we are concerned about the unrepaired or irreversible injury, and we know that occurs and that's immediate.

So the cell, the person has the injury right away. It's not a question of the injury developing later. He has been injured the moment the cell has been irradiated.

Thereafter the only way you detect a cancer is when there is enough of it to be felt or seen or detected by an X-ray examination as a spot on a lung. Now, that takes, generally speaking, an amount of cancer of at least of the order of about one gram, about a four-hundredth of a pound, and when you have one gram of cancer, you already have about a billion cells.

So, when you ask me how long does the cancer take to develop, in essence the cell that is predestined to cancer from the injury it received immediately is there right at the time of the radiation.

How long it takes for one cell to divide to two and two to divide to four and even further changes in that cell, as it goes on to become that billion or more that it takes to be detectable, can be very variable.

We can see in experimental animals cancers in months or less than a year.

In the human, certain forms of leukemia have been proven as early as three to five years after the irradiation.

Most of the solid forms of cancer, when we think of lung cancer, kidney cancer, brain cancer, colon cancer, breast cancer, we think of periods more like ten, fifteen, twenty years.

But I would like to point out that it's a fallacy to think that nothing is happening between the initial injury and ten years.

When we say that we see the cancer provably at ten years, it means for the number of people that have been studied in a given observation, that it was only possible to prove it definitively at ten years. If you had a hundred times as many people, you might have proved the cancer's existence in two years or even one year.

Q. ***Can leukemia or cancer be specifically identified as caused by ionizing radiation?***

A. There is no reasonable doubt in my mind or to my knowledge from the scientific literature on the part of anyone that radiation is a cause of leukemia or cancer.

Now, the way you asked the question, I believe, is can it be identified?

A specific cancer or leukemia does not raise a little flag which indicates that radiation was the causation, and since there are other causes of cancer besides radiation, we cannot specify that a given cancer was totally caused by radiation.

But the evidence beyond any reasonable doubt is if you take any two groups of humans, otherwise identical, irradiate one group and not irradiate the other, there will be provably more cancers and leukemias in the irradiated group than in the non-irradiated group.

And, moreover, if you have a human subset,

as for example occurred in Hiroshima and Nagasaki survivors, where there were people who were at various distances from the radiation source, mainly the radiation from the nuclear weapon, we can subdivide those people into those who had successively higher amounts of radiation and the number of cancers goes up the more the radiations. So, that is the nature of the proof, and it is beyond a reasonable doubt.

Q. ***What is the effect of radiation on the developing fetus?***

A. Radiation injures, as I mentioned earlier, the genetic material, and it's that material that is guiding the cells in a developing fetus to form the various organs and tissues that have to be formed to go from an ovum all the way to an embryo and finally to a fully developed human. There is considerable evidence in a variety of types of studies which indicate that the developing fetus is more sensitive to ionizing radiation in terms of the effects caused than are children, and children more sensitive than adults, and even within the developing fetus in the first trimester of pregnancy, the fetus is much more sensitive to radiation injury than in the third trimester of pregnancy.

Q. ***So what would be the result of this effect?***

A. The result we have now of studies confirmed around the world, based upon very large population samples initiated over twenty years ago by Dr. Alice Stewart in Great Britain and which have proved beyond statistical doubt that for fetuses irradiated just by a very small dose of diagnostic X-rays in the third trimester of pregnancy of about a fifty per cent increase in the incidence of cancer, cancer fatalities of all types, leukemia of all types, during the first ten years of life, just from the amount of radiation received from a diagnostic study of the mother; and for fetuses in the first trimester, the sensitivity is something on the order of ten to fifteen times as high.

And, moreover, Dr. Stewart's work has shown by comparing women that had, just for a variety of reasons but not related to their health, one, two, three, four or five X-ray films during that examination, that the number of cancers and leukemia in the children of those women goes up in proportion to the amount of radiation the woman had, all at very low doses of total amount of radiation.

Q. ***What is the result of radiation in the overall population and/or the gene pool?***

A. And/or the gene pool, did you say?

Q. ***Yeah, the overall population and the gene pool?***

A. There are—first of all, if we are talking about a massive irradiation dose in terms of the unit that is usually used, rem or roentgen or the rad, when you're talking about irradiation doses in the neighborhood of three hundred to four hundred rems, if you do this all at one sitting in a fraction of a second or a few minutes or an hour, you would kill fifty per cent of the people outright.

But the concerns, I think we are discussing primarily in the nuclear fuel cycle or what we refer to as the effects of radiation at more modest doses, and the effects of radiation at those more modest and even low doses is very serious, but it's of a different type, not immediate type deaths, but rather the occurrence of extra cases of death due to leukemia, the occurrence of extra cases of death due to every major form of cancer.

And for those people who are still in their reproductive years, either male or female, we have the injury that can occur either to the sperm-generating cells in the testes and to the ovum-generating cells in the ovary, and injury to the genes there can provoke hereditary changes and diseases and deaths in generations for many generations beyond that of the irradiated individuals.

So we refer to two types of injury due to low and modest doses of radiation; somatic, meaning those that occur in this generation, and those are cancer and leukemia; and genetic, meaning those effects that will occur in subsequent generations as a result of the irradiation of this generation.

Q. ***Could you briefly explain how the nuclear fuel cycle works?***

A. The nuclear fuel cycle is essentially a system devised to extract some of the energy that is potentially available from such substances as uranium or plutonium or thorium.

In the case of the nuclear fuel cycle currently in place, we are using one of the forms of the uranium that occurs in nature, the so-called uranium 235 isotope.

This in nature is only seven-tenths of a per cent of uranium, and in the fuel cycle in place in the United States, we must first enrich that uranium to about two to four per cent uranium 235.

So the steps you have are first to somewhere find the source of ore that is rich enough in uranium to be worth extracting, namely the money and energy costs of getting uranium out being such that you think you can get more energy back.

You must isolate the uranium from that ore and leave over the mountain of residues that are radioactive from that ore.

Then in the United States cycle in place now, you must go through, at this moment, the process mostly used is gaseous diffusion, to separate the lighter uranium 235 from the 238 and thereby enrich it, and then place it in a device known as a reactor, which is a configuration in which, if your arrangement is correct, you can get uranium 235 nuclei to undergo fission.

And in undergoing fission, they produce neutrons, the same as the neutrons that initiated the fission.

And if you have things arranged properly, there are enough neutrons left over after everything else that might steal one of the neutrons in the reactor to keep the chain reaction going.

And for every uranium 235 that we fission, we get about two hundred million electron volts of energy, and that's a large amount of energy per nucleus, and will produce radioactive by-products called fissionable products, which have long, long half-lives and will have to be isolated.

At the same time, from some of the major components in the nuclear fuel cycle, mainly the uranium 238, we produce the by-product known as plutonium 239 through the capture of some of those neutrons and a couple of intermediate reactions.

Now, this chain reaction produces the energy, as I mentioned by the fissioning.

Water is circulated through the clad tubes that contain the uranium fuel. That water is cooling the fuel and itself being heated, and either is kept under pressure and brought on to a high temperature by the chain reaction, and the water is transferred through a series of pipes to a steam generator where it transfers its energy to a secondary set of pipes and then gives rise to the so-called pressurized water reactor, or the water that is passing by the fuel elements can go directly to a turbine and turn the turbine.

In the former case, the pressurized water, it is the water in the secondary system.

And so you finally have your heat converted either into hot water that transfers to a secondary system and then steam, or you have steam produced directly by boiling water, and you turn a turbine and generate electricity from this energy.

Most of the energy is lost, of course, because we cannot transform more than a certain amount of the energy as heat into electricity. That's limited by the laws of thermodynamics.

And, unfortunately, the accumulation of some of the by-products of the reaction, namely, the fission products, leads to an accumulation of materials which have a tendency to steal the neutrons, so that every year in a three-year cycle you must take some of the so-called spent fuel rods out of the reactor and replace them with new fresh uranium; and these spent fuel rods have a very, very high radioactivity in a modern thousand megawatt power generating reactor,

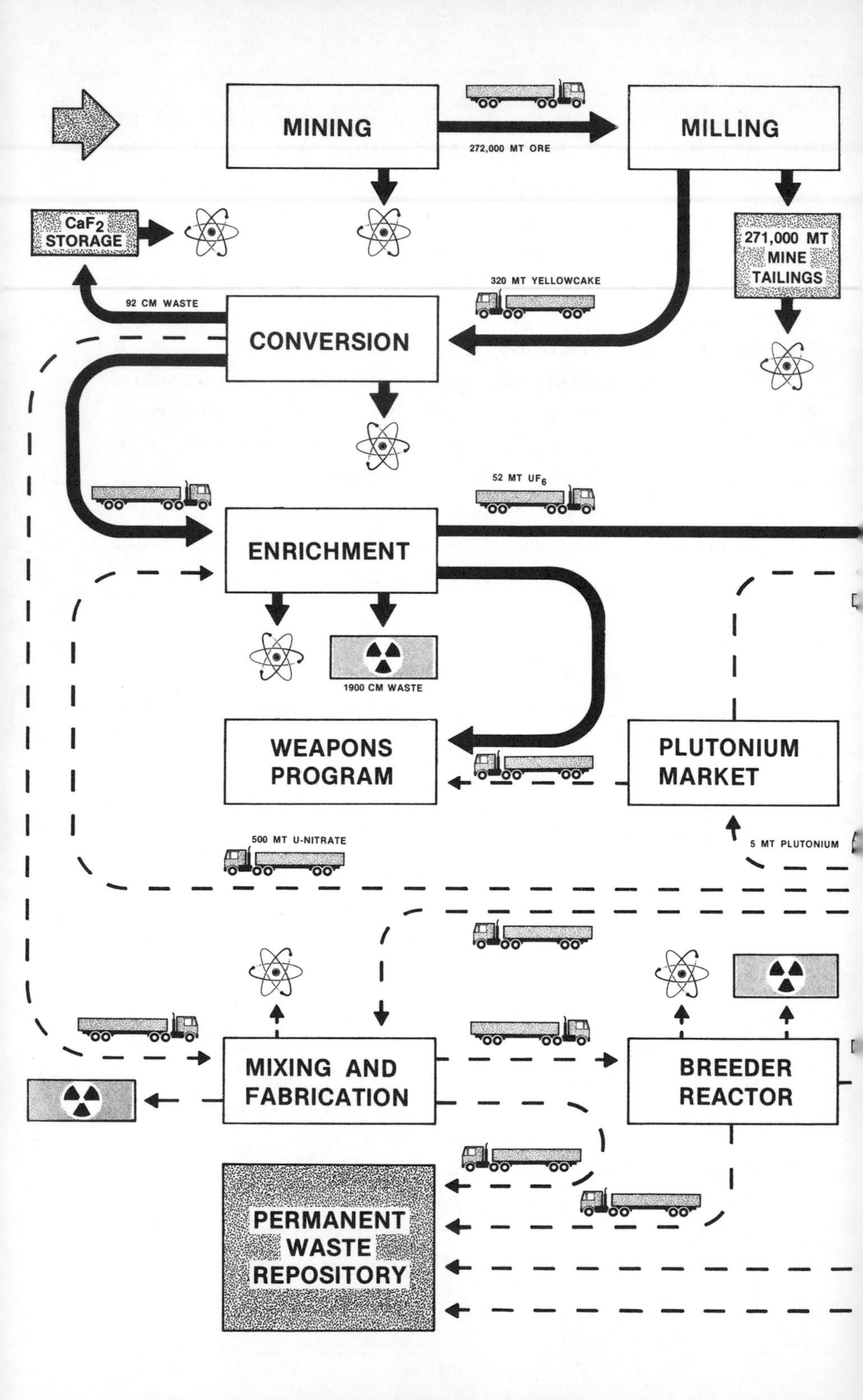
MINING
272,000 MT ORE
MILLING
CaF$_2$ STORAGE
271,000 MT MINE TAILINGS
320 MT YELLOWCAKE
92 CM WASTE
CONVERSION
52 MT UF$_6$
ENRICHMENT
1900 CM WASTE
WEAPONS PROGRAM
PLUTONIUM MARKET
500 MT U-NITRATE
5 MT PLUTONIUM
MIXING AND FABRICATION
BREEDER REACTOR
PERMANENT WASTE REPOSITORY

THE NUCLEAR FUEL CYCLE

A ONE YEAR SUPPLY OF URANIUM FOR A SINGLE REACTOR BEGINS AS 272,000 TONS OF RAW ORE, AND TRAVELS OVER 100,000 SHIPPING MILES BEFORE BECOMING 70 METRIC TONS OF HIGHLY IRRADIATED WASTE, AND MORE THAN 300,000 METRIC TONS OF LONG-LIVED WASTES.

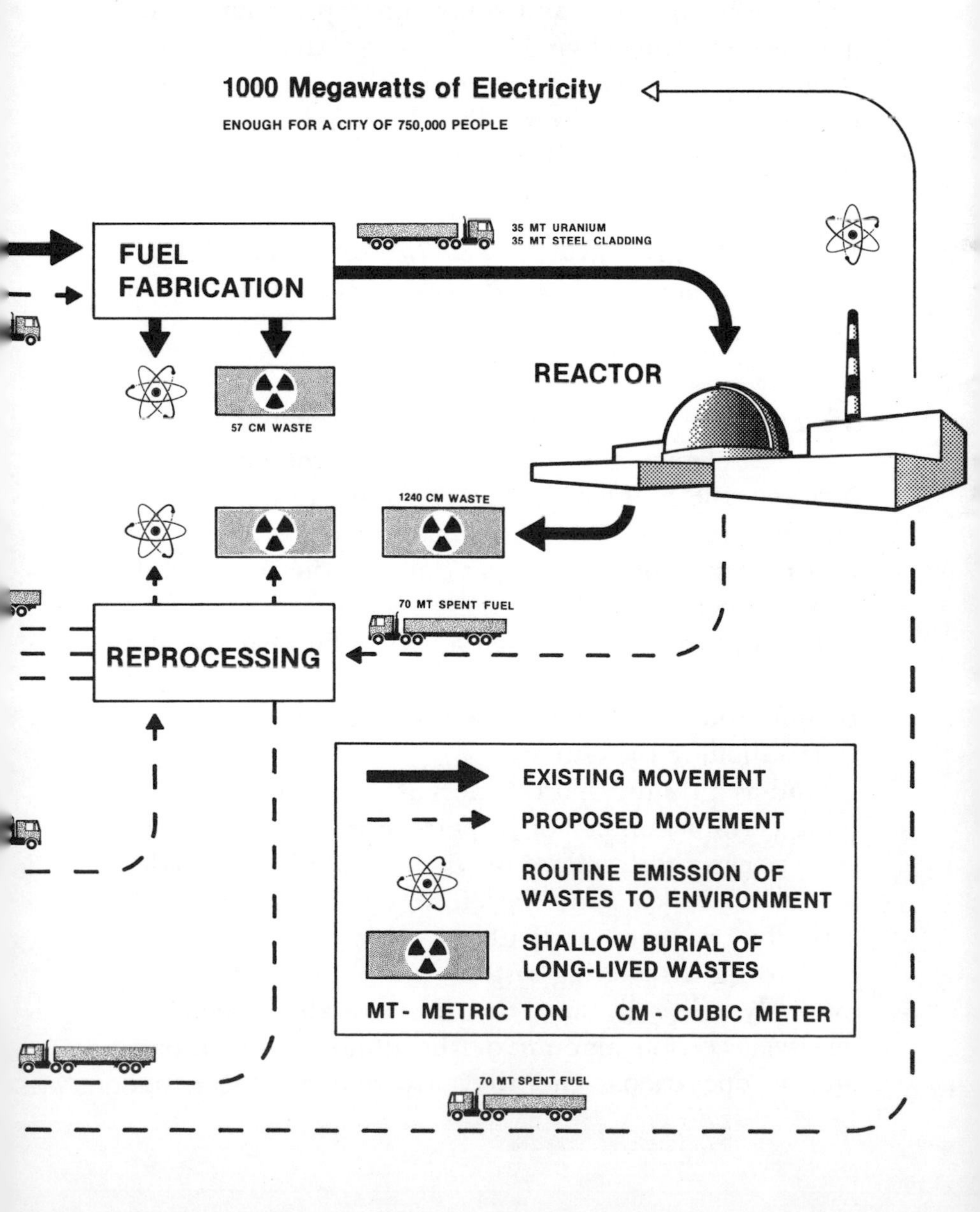

and they must be in some way isolated from the biosphere.

Either they are kept for the moment in the form of the spent fuel rods, or, in some experimental reactors and in some other countries, the fuel rods are chopped up and dissolved and then an effort is made to recover the unused uranium and some other plutonium that has been produced and to separate that from the fission products which are regarded as a waste that must be isolated from the biosphere essentially indefinitely and virtually perfectly.

Q. ***Is the routine release of radionuclides inherent in the design of the nuclear fuel cycle?***

A. Yes, the cladding that is put around the fuel elements in a reactor can never be perfect. You see, the problem is that you want to have a thin cladding to keep most of the radionuclides inside the fuel rod and not have it leak out into the water while you are operating.

But also you need to have a very good heat transfer between the water and the fuel rod in order to prevent the fuel rods from melting, which is a thing to have to worry about in the systems.

And as a result, that thinness of the cladding makes for a certain number of imperfections, and we know that a certain fraction of fuel rods have pinholes and other leaks, and, therefore, some of the radioactive by-products, particularly noble gases and iodine get into the circulating system, and some of them are routinely released, and the licenses allow them to release a certain amount of the radioactivity during routine operations, and they are released by actual

Boiling Water Nuclear Power Reactor

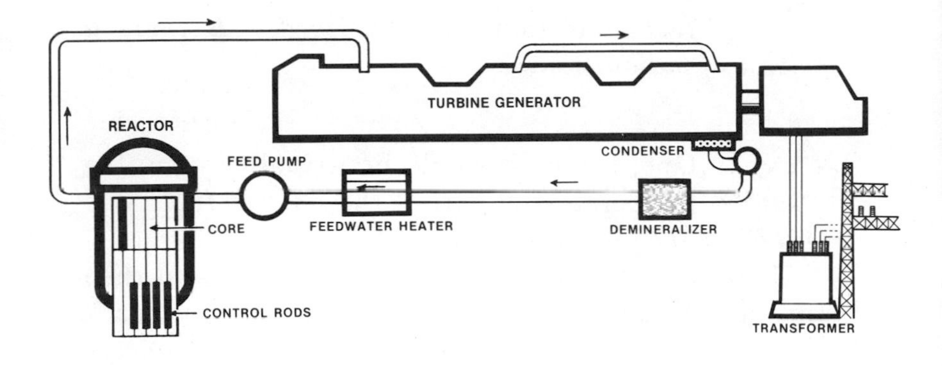

Pressurized Water Nuclear Power Reactor

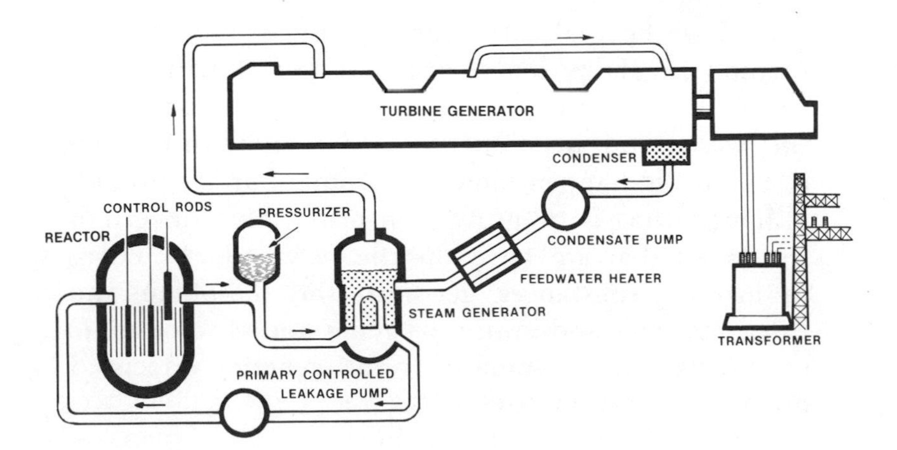

measurement from all of the nuclear power plants that are operating.

Q. ***What about the other parts of the nuclear fuel cycle that you described? Is there a release of radionuclides?***

A. One, I think it would be virtually impossible to operate any part of the nuclear fuel cycle without some release of radionuclides.

If you were to chop up the fuel elements and then try to dissolve them so you could get back your uranium and plutonium, you would release possibly even greater amounts by far than in the nuclear reactor itself, and that's a concern for some of the radionuclides, such as carbon 14, a very dangerous one and probably one of more concern with respect to health effects than many others. There's no technology even known for keeping that from getting out at the present moment in any part of the fuel cycle.

And we hear about waste disposal. We have no method in place, and I think a fair description is we really have no method in mind for disposing of the waste and my concern is far more not what is going to happen down in some burial ground, although that is of grave concern. My concern is the losses that we have on the way, which means radioactive substances get into the biosphere and thence to man and injure his genes and chromosomes producing those somatic and genetic effects I mentioned earlier.

Q. ***Okay, you mentioned plutonium earlier. Could you tell us about plutonium?***

A. Yes, I mentioned earlier that plutonium is not a fission product.

It is a by-product of the operation of the reactor and is regarded by those in the nuclear industry as a valuable by-product, because plutonium itself, plutonium 239, one form of plutonium—there are several—is made from uranium 238; that's not the usual one used in the reactor to generate most of the fission energy, but the plutonium 239 is made. It can be re-isolated and will help support a chain reaction in a new reactor, because you can split it.

The problem with plutonium is that the form you would want to use it in a new reactor is in the form of plutonium oxide, which is one chemical form, and in order to prepare that you get it into fine particles of plutonium oxide about a millionth of a yard long. They are tiny particles, and they are precisely the type of particles which, if inhaled by men, will lodge in the lung, and we have estimates of the lung cancer potential of those plutonium particulates. We regard it as one of the most serious lung cancer agents that we could think of, because it emits alpha particles, and alpha particles, energy amount for energy amount, are expected, from all of the data we have today, to be about ten to twenty times as effective as other forms of energy.

Now, there's not a shred of doubt that alpha particles produce lung cancer in man. We have that evidence from the uranium miners, and the alpha particles from plutonium are just about the same energy as those in the uranium miners' exposure.

So, there's no doubt.

There is some controversy among workers in the field as to exactly how long these particles of plutonium would lodge in the lung and how long they lodged determined how many cancers you will get.

My estimate is for the kind of plutonium that you get from the nuclear fuel cycle, that in nonsmokers of cigarettes, about a four hundredth millionth of a pound will guarantee human lung cancer, or stated another way, a pound of plutonium has enough in it if finely divided and put into human lungs to cause four hundred million human lung cancers.

And my estimate is that the cigarette smokers in the population, because of the damage to part of their clearance mechanism, their lungs might be a hundred times more sensitive to the effects of plutonium.

If the fuel cycle does initiate the step of reprocessing to get back this plutonium—we have done it experimentally—and there was a period for a while in New York where a company did do some reprocessing of commercial fuel; that is closed down now—we will be handling thousands and thousands of pounds of plutonium in the fuel cycle, and that is where the hazard comes up, because of its enormous lung cancer potential.

That's not the only effect of plutonium. If it gets into the biosphere, on the ground and into waters, some of the plutonium is fairly insoluble and not easily taken up by.plants or by man through eating it. As I mentioned, the hazard through inhalation is enormous.

But the recent works on plutonium indicate it's even a lot more of concern by ingestion, that is, eating, than was thought before, because plutonium has the notorious capability of interacting to form very, very tight chemical complexes with certain molecules, organic molecules, that occur in nature, and these things can facilitate the uptake of plutonium into plants and, hence, into man.

And rather recently there has been scientific evidence that shows that all the estimates of the low hazard of any plutonium—that has nothing to do with the very high hazard of breathing it—that the low hazards estimated about eating it are wrong and wrong by about a thousand times, because plutonium in the presence of, for example, drinking water that has been treated with chlorination, which is the case in the United States very widespread, gets converted from the plus three or plus four oxidation, or what we call valence state to plus six, and the plus six state is very much more readily absorbed than is plus four.

I have worked with plus four and plus six plutonium in the laboratory myself, and I know that the behavior of plutonium in the plus six state is very, very similar to that of uranium.

In fact, I developed and patented a process for separating plutonium based upon this.

And so this grave error in the underestimate of a hazard from ingestion may even make the eating of plutonium as bad a problem as the breathing.

Q. ***What are the chances of plutonium escaping?***

A. Well, I think that's a question that's really related to Murphy's Law in some ways, which states that anything that can happen will.

I could not say what the chance is in a given operation that one per cent, a tenth of a per cent or a hundredth of a per cent will escape, but I can give you experience.

For example, plutonium is handled in great amounts at Rocky Flats, Colorado, and in 1969 they had a big fire there, and there was some concern about how much plutonium had gotten out.

And the Atomic Energy reports were that no more than a milligram of plutonium had gotten out.

Dr. Edward Martell was skeptical about that and went out and measured plutonium outside the plant miles away from the plant and found plutonium on the ground there. In fact, the final estimate he made was about half a pound, which indicated that the Atomic Energy Commission had underestimated the escape of plutonium by two hundred thousand times.

The Atomic Energy Commission then put their own research workers from the Health and Safety Laboratory of New York onto the problem, and they confirmed the general size of Martell's estimate of how much had escaped.

Later it turned out that it wasn't the fire that had caused the escape of plutonium, but it indicates a pathway that they hadn't even thought of before for much more escaping than the engineering calculations would have led them to believe; and this is why the Atomic Energy Commission was so gravely in error on the amount they had released.

What was the mechanism? They were machining plutonium in this plant, and they had some plutonium left over from the machining, and they would be putting them in barrels of oil to store until some later date that they would be able to reprocess and get the plutonium back. But momentarily in the Rocky Flats plant they didn't have the facility to do that, so they stored the barrels in an area called the 903 Area at Rocky Flats out in the open.

And finally they had approximately five thousand five hundred barrels stored there, and barrels have a notorious capacity of rusting, and about a fourth of the barrels did rust out, and the plutonium and all leaked into the ground, and then the plutonium was available as particles on the ground, attached to soil particles, and that area is characterized sometimes by winds in the forty to seventy mile an hour region, and the winds picked up the plutonium particles, and that's how half a pound at least got moved off site nearly all the way to Denver.

So in answer to your question, I feel I have answered it, that all kinds of things can happen that can lead an engineering calculation that no more than one per cent or a tenth of a per cent or a thousandth of a per cent will get to get out to be off by two hundred thousand times.

Q. ***And how long does plutonium last in the biosphere?***

A. Plutonium has a half-life—plutonium, the 239, the major form, has a half-life of twenty four thousand four hundred years.

Now, there's a very simple way to decide how long something is hazardous biologically.

You see, in one half-life there is only half as much of any amount you start with. In two half-lives, there's a fourth.

And scientists generally feel unless you started with an astronomical quantity, that if you wait ten or twenty half-lives, you are down to a fairly innocuous situation.

So for plutonium, you would only have to wait two hundred forty thousand to four hundred eighty thousand years to have it rendered fairly innocuous.

Q. ***Okay. You have worked on a number of government projects since the earliest period of the country's experimentation with nuclear power. Could you please briefly explain the history and development of standards of safety for the industry?***

A. Yes, I have been with nuclear energy since before there was an Atomic Energy Commission and before there was a Manhattan Project. In fact, I was a member of the early team that did some of the work that led up to the Manhattan Project.

At that time we knew that radiation was harmful, because we had the experiences of the early workers with X-ray and with uranium and radium, including Madame Curie, who died from radiation poisoning. So there was no doubt of the harm of radiation.

But for a very strange reason, and based upon no evidence at all, some scientists made the presumption that just because you could recover from the acute effects of a dose of radiation, say, you gave a lot of radiation to the skin and the skin became red, and then the reddening went away, they made the incorrect assumption that the danger of cancer or leukemia might be also correctable by spacing the radiation out. No one really believes that any more, but at the time I started to work in the Manhattan Project, there were such notions kicking around, and so the amount permitted people was quite high.

In fact, in 1954, which was early history, but still fourteen years beyond the time I started to work, there's an organization, quasi-governmental, ostensibly concerned with radiation protection known as the National Council on Radiation Protection and Measurements, and they issued a statement that we could give one-tenth of a unit per day to people without a physical effect, thirty-six units per year.

I might point out to you that today and in force for the last ten years is a statement that we can't even give one-fifth of a unit per year. In other words, the standards have come down a hundred and eighty times in the period from '54 to '69 in terms of what people thought it was all right for humans to take.

And in 1979, by recommendation of the Environmental Protection Agency, a standard approximately a sixth of that onc is going into effect.

So you have had an enormous tightening of the standard, because the history of all of the things that have come to light from people irradiated in a wide variety of circumstances, that radiation is a far, far more potent cancer producer than was thought, and leukemia producer.

And from the genetic evidence that developed during this period of the Manhattan Project up through the Atomic Energy Commission, it was proven that it wasn't a matter of high doses; that as you went down in dose, you got lesser effects, but there was no evidence of a safe dose at all.

And so every one that I know of, every international body, the literature of which I read, concerned with radiation protection, operates on the principle that there is no such thing as a safe dose, and certainly no scientific evidence has ever been adduced that there is any amount that is safe.

Q. ***At this present time would you still call nuclear power an experiment on human...***

A. Nuclear power is one of the greatest experiments being conducted on the human species and on the biosphere in general. It's sort of a game of chance.

We create—no one can argue, it's just an arithmetic exercise to figure out the quantity of radioactivity we produce, and the name of this whole experiment is, when you consider the enormous complexity of the fuel cycle, all of the steps that must be gone through, chemically, transportation and otherwise, everything hinges on what percentage of this great quantity of radioactivity gets out.

If the percentage is a trillionth of the amount you make, only a small number of people would die.

If the percentage is a thousandth, we will have a cancer disaster of great magnitude and a genetic disaster.

So, it's a question of what we'll experience over the years under all circumstances of acts of God, tornadoes, malevolence, human error, machine malfunction, what will be the percentage that gets out? That's an experiment, and we don't know that answer.

Q. ***All right, I would like to introduce—show you Exhibits 6 and 7 and ask you what they are? Could you tell us?***

A. Which one?

Q. *That one first, the tables first.*

A. Yes, I have seen this table before in connection with some of my studies and participation in the GESMO, that's the Generic Environmental Statement for Mixed Oxide Fuels, and this is the Final Environmental Statement on Plutonium Recycling in that same set of hearings, and I have studied that, too.

Q. *And what do those tables mean?*

A. These tables describe the estimated health effects.

Health effects is a word that is commonly used in the industry to mean deaths from the light water industry in the period from 1975 to 2000, and they describe the expected number of cancer deaths from bone cancer, thyroid cancer, lung cancer and all forms of cancer, and they describe it, the genetic defects, certain types, and all genetic defects if we just stored our fuel rods without doing anything with them, if we do chemically chop them up and break them and only recover the uranium, which is called the uranium recycle in this table, or if we chop them up and try to recover and reuse the uranium and the plutonium.

My personal scientific work and analysis of this data indicates that the Nuclear Regulatory Commission has underestimated the effects of all these things by a very large factor.

TABLE S-5

Estimated Health Effects from U.S. LWR Industry 1975-2000*

Source: *NUREG-0002, Generic Environmental Statement on the Use of Mixed Oxide Fuels in the LWR Fuel Cycle* (*GESMO*), *NRC, 1976*

	No Recycle				*Option U Recycle*				*U + Pu Recycle*			
Type of Health Effect	Occ.	U.S. Non Occ.	Foreign	Option Total	Occ.	U.S. Non Occ.	Foreign	Option Total	Occ.	U.S. Non Occ.	Foreign	Option Total
Bone Cancer Deaths	45	90	6.9	140	42	97	23	160	39	90	22	150
Benign and Malignant Thyroid Nodules	1,300	160	69	1,500	1,230	800	300	2,300	1,200	800	300	2,300
Thyroid Cancer Deaths	51	6.6	2.8	60	50	32	12	94	48	32	12	92
Lung Cancer Deaths	360	31	4.7	390	330	53	29	420	290	51	27	370
Total Cancer Deaths	**550**	**530**	**28**	**1,100**	**540**	**620**	**120**	**1,300**	**530**	**570**	**120**	**1,200**
Specific Genetic Defects	650	620	33	1,300	630	730	140	1,500	620	660	140	1,400
Defects with Complex Etiology	410	390	21	820	400	460	91	950	390	420	89	900
Total Genetic Defects	**1,100**	**1,000**	**54**	**2,100**	**1,000**	**1,400**	**170**	**2,400**	**1,000**	**1,100**	**230**	**2,300**

*Exposed populations are indicated as follows: *Occ.* = occupational exposure of U.S. LWR industry worker; *U.S. Non Occ.* = Non-occupational exposure of the United States population; *Foreign* = nonoccupational exposure of world population, excluding U.S.

Q. ***Does the Nuclear Regulatory Commission currently use these tables?***

A. Well, I think this is dated something like 1976, and I'm not sure what their thinking is as of today in 1978. I'm sure they are constantly undergoing revisions in their own thinking.

I do not have their up-to-date view on every number in this table.

Q. ***Okay, this report is called the GESMO Report?***

A. Yes, these are both parts of the general proceedings that were called, the preparation of a Generic Environmental Statement on Mixed Oxide Fuels. At the time the United States was considering taking the spent fuel assemblies from the reactor, chemically dissolving them and separating out and getting back the plutonium and uranium, and then making new fuel rods partially out of plutonium and partially out of uranium, and this whole set of hearings was to consider all of the kinds of impacts, economic, health, and other, of doing that. These hearings were suspended by the decision, as pursuant to the decision of the President of the United States, to not go ahead with recycling at this time.

BY MR. KACHINSKY: I move to have these exhibits placed in evidence.

BY THE COURT: Without objection Exhibits 6 and 7 are now in evidence.

[*Marked and filed Exhibit Nos. 6 and 7 in evidence*]

BY THE COURT: This is a good time to go to lunch. It's 12:00 o'clock. We will recess until 1:00 o'clock.

BY THE CLERK: Everyone rise, please, court is in recess until 1:00 o'clock.

[*Thereupon Court recessed for lunch*]

John Gofman and Albert Bates

BY MR. KACHINSKY: [*Continuing*]

Q. ***I would like to ask you, Dr. Gofman, do these tables also, besides indicating the***

recycle option, indicate the present fuel cycle presently being used?

A. Yes. This table is divided into three parts. The no recycle is what we call the throwaway cycle of just keeping the spent fuel rods that we take out of the reactor; that is the current system, and the other two columns of the table are possible future options.

Q. ***And what is the number of deaths assessed for that present fuel cycle option?***

A. For the present fuel cycle, that is, the way we are going now, and the estimate would be for the current industry, between 1975 to 2000, would be one thousand one hundred total deaths from cancer in the U.S.A.

Q. ***And what about for genetic...***

A. I'm sorry, let me correct that. Of that one thousand one hundred deaths, twenty-eight of these would be in foreign countries.

And what about the genetics?

A. The genetic defects estimated in this table would be for this same cycle that we are now in, two thousand one hundred genetic defects.

Q. ***Okay, thank you. Do you know Alice Stewart?***

A. Yes, I do know Alice Stewart.

Q. ***Could you give a brief summary of her credentials?***

A. Dr. Alice Stewart is one of the world's most renowned radiation epidemiologists, and she has now to her credit two major landmark studies that are very relevant to the nuclear fuel cycle.

The first, the demonstration which I alluded to earlier today, that very low doses of radiation, those in the diagnostic X-ray range, can produce a fifty per cent increase in all forms of cancer and leukemia in children when the mothers are irradiated, and more recently Dr. Stewart has collaborated with Dr. Mancuso and Dr. Kneale in producing a report showing the extra cancer deaths in atomic workers at the Hanford plant at the doses of radiation that are allowable, which is a study which shows that the allowable dose has nothing to do with being a safe dose.

Q. ***Okay, did she do a study on leukemia?***

A. The study she did on the children was a study both on leukemia and on various forms of cancer, and the results for leukemia were just about the same as the results for other cancers, approximately a forty to sixty per cent increase associated with just X-raying the mother.

Q. ***Were efforts made to discredit this study?***

A. Oh, yes, this was a study that went down hard with people who had this myth in mind that you had to have a high dose of radiation in order to get cancer and leukemia.

These studies were initially reported in 1956, '58, and there were all kinds of statements that it couldn't be so and it shouldn't be so, but it has been confirmed now in several studies.

Hers were done in Great Britain, vast numbers of individuals in the studies, not a question of statistical problems. She had way more than enough to prove her point.

It has been confirmed by a study by MacMahon in this country. It has also been confirmed in a study by Gibson, the so-called Tri-State study, and in two successive years, two leading British radiation biologists, health physicists and radiologists, one Dr. Robert Mole has confirmed that he believes Alice Stewart's data proves causation of cancer and leukemia by radiation in those children.

And a year later in the British Journal of Radiology, Dr. Eric Pochin has also confirmed it.

So I know very few people in the scientific community now who take this issue with Dr. Alice Stewart's studies.

Q. ***Have you also analyzed her data independently?***

A. I have analyzed independently the study of the atomic workers. I have not done an independent study of the children.

But I have done an independent study totally from scratch using the prime data for the Hanford workers in which it was proved by Alice Stewart and Dr. Mancuso that the so-called occupational dose is not safe, and I have reached my own conclusions by a method of approach to the data totally different, a method that I believe will withstand any of the criticisms that have been levelled at some of the previous analyses, and I arrived at essentially the same conclusions that Dr. Stewart and Dr. Mancuso do.

Q. ***And what is that conclusion?***

A. That conclusion is they said it takes about thirty-three units of radiation, the rem unit, to double the frequency of cancer in workers.

In other words, if the workers were going to get ten cancers for every thirty-three rems, they will get another ten, double what the spontaneous is. That is the Stewart-Mancuso conclusion.

My conclusion is thirty-eight rems to do the same, and that, incidentally, since I have published such analyses before from people getting higher doses of radiation, indicates that I had previously underestimated the hazard of radiation. The true cancer hazard is worse than I thought, and my number is thirty-eight, theirs is thirty-three—to double—between thirty-eight and thirty-three arrived at by totally independent methods of analysis consists of excellent agreement; ten per cent apart is nothing in such studies.

Q. ***How has this study been received by the Nuclear Regulatory Commission?***

A. My study or Dr. Mancuso's?

Q. ***Dr. Mancuso's.***

A. The material I have read from the staff of the Nuclear Regulatory Commission is that they have acknowledged the study.

They have said that some people have criticized the study as having methodological deficiencies and ambiguities.

But they have also studied in the material from the Nuclear Regulatory Commission that I received from them that there appears to be something there, and I believe my own analysis is free of any methodological ambiguities.

Q. ***Okay, are there any other instances that you know personally that scientists have been discredited?***

A. You mean the effort to say their studies were wrong?

Q. ***Yeah.***

A. I know quite a number, the Atomic Energy Commission during its existence did not look favorably upon people who indicated that radiation was harmful, because their whole thrust was to suggest that there was some safe dose of radiation.

Dr. Arthur Tamplin, my colleague, when we published the paper showing that radiation would produce twenty times as many cancers per unit of radiation as had been thought, Dr. Tamplin had some of his scientific papers censored.

He had twelve of his thirteen scientific colleagues taken away from him.

He was honored by the American Cancer Society with an invitation to come and talk to them about his work, and ordinarily in our laboratory sponsored by the Atomic Energy Commission that's a great feather in the cap of the laboratory, but Dr. Tamplin had two days pay docked to go to that meeting, which had never happened before to a scientist, to my knowledge.

Myself, my staff was not taken away in connection with this, the radiation work, but I did lose two hundred fifty thousand dollars a year from my cancer chromosome work, which is directly a harassment for my position on this.

More recently when Dr. Mancuso made his announcement of his findings that the radiation was more harmful, was even more harmful than I said it was eight years ago, he had his funds cut off.

Congressional hearings indicated that there was no justification for cutting those funds off. That's in the Congressional Record now.

Dr. Irwin Bross is another illustration of a man who has stated that radiation is more harmful than people had thought, and, indeed, that there are some people that are many, many more times susceptible to radiation injury than the average, a point that is extremely important, because all thinking in radiation is that you should worry about the most susceptible people.

Dr. Bross' funds have been cut off by the National Cancer Institute. There have been, I would say, very serious criticisms of Dr. Sternglass's work in attempts to discredit—there's a pattern, in my opinion, of harassment and efforts to discredit scientists who find radiation more harmful.

Another illustration is a former colleague of mine, Dr. Donald Geesaman, who at that time was one of the few people in the Atomic Energy Commission who was doing work on the lung cancer hazard of plutonium, and he was dismissed ostensibly as a reduction in work force. I just mention the fact that he thought plutonium was more hazardous than others before him. He lost his position.

I think that's a fair summary of those cases I know of by direct experience and knowledge.

Q. ***Have you done extensive work on the induction of cancer by certain radioactive elements?***

A. That has been the major field of endeavor I have been in over the past fifteen years, and more particularly in the past ten I have worked on estimating the number of cancers from irradiation in general from specific fission products, and in particular I have done a great deal of research on the induction of cancer by plutonium 239.

Q. ***Could you tell us about the study you did that you published indicating that a large***

number of cancer and birth defects would occur if the permissible dosage of radiation actually was released?

A. That is a very simple statement. There is and has been in force originally from the Federal Radiation Council that the average person in the United States shall be permitted to receive .17 units of radiation per year.

Now, the Federal Radiation Council never said that this amount of radiation is safe. They didn't say it was without harm, although many people have misinterpreted that. I think the Atomic Energy Commission in its existence, for example, through its chairman, when he got on national CBS television said, 'I think there is a safe dose.' So, that was in existence as an allowable dose.

And since I had been hired by the Atomic Energy Commission to work at the Lawrence Livermore Laboratory to find out just how much cancer and leukemia and genetic injury could be produced by radiation in 1969, when we had numbers available, Dr. Tamplin and I presented those numbers at a very serious scientific meeting of the Institute of Electrical and Electronic Engineers; that initially we said sixteen thousand cancer deaths would occur, cancer plus leukemia deaths would occur if everybody in the United States got on the average the permissible dose.

We subsequently—we said we thought that number might be too low, and with further work we raised it to thirty-two thousand as our best estimate.

As a result of our work, Mr. Robert Finch, Secretary of HEW—we had already testified before a Senate Committee, but Senator Muskie asked Robert Finch,

Secretary of HEW, what he was going to do about it, and Mr. Finch asked the National Academy of Sciences to study the question.

They appointed a committee known as the Biological Effects of Ionizing Radiation Committee, and after a two-year study, they published an extensive report, and they suggested that my number might be five times too high, but I would consider and so the scientific community considered that that wasn't all that much disagreement. We regarded that as essentially a vindication of our position.

We now know from the studies of the Hanford workers that I was two or three times too low in estimating the hazard, and the BEIR Committee far, far too low.

Q. ***Is the BEIR Commission still accepted for 1972?***

A. Well, I have pointed out in writing that its estimates are too low and why I think they are too low, and Dr. Radford, the current chairman of the BEIR Committee in testimony before the Congress in February has stated that the BEIR Committee is going to raise their estimates, because the hazard now appears to be worse than he thought, which is in the right direction.

And we understand that there will be a report from the BEIR Committee this fall sometime revising their previous estimates.

Q. ***Could you tell us what ALARA, A-L-A-R-A, refers to?***

A. ALARA, A-L-A-R-A means as low as reasonably achievable. It has nothing to do with safety or freedom from cancer and genetic injury. It just means that for the amount of money you are willing to spend, try to do what you can to keep people from getting too much of a dose and hence too many cancers and leukemias and genetic injuries.

Q. ***Does ALARA essentially plan in human deaths?***

A. It permits deaths.

Q. ***Permits human deaths?***

A. Yes, because ALARA does not say—see, the only way you could avoid deaths from the nuclear fuel cycle is to have zero releases.

ALARA says keep the releases as low as you can reasonably achieve with the economics that you want to spend on it and the equipment you have available and so forth.

So it is a planned emission of radioactivity and that in effect means planned deaths.

Q. ***What is a lifelong plateau?***

A. The effect of radiation in producing cancer is that for a period and after the person is exposed—he's injured the moment he's exposed; that's when the genes and chromosomes are hurt; that's an irreversible injury.

But then at some time later, as I indicated before, you begin to be able to perceive an excess of cancer.

Now, in the human, the studies that have been going have not been going long enough so that we know for sure whether once you start seeing, say, a thirty per cent increase in cancer per year, whether it will last for twenty years or thirty years or for the whole rest of the life of the persons exposed. We call this region where the number of cases of cancer deaths each year caused by the radiation, where that number stays fairly constant, we call that a plateau region.

And the question asked in your question about a lifelong plateau is the radiation effect continued throughout the rest of the life or do you go back to the normal risk after thirty years or so?

Everyone in the radiation community of protection knows that we don't know the answer, and the only reasonable prudent public health posture is to assume a lifelong plateau unless you have proved that it isn't lifelong.

But the number of cancer deaths is much higher if the plateau lasts for the lifetime than if it only lasts for twenty or thirty years.

Q. ***What was the Tri-State Leukemia Study?***

A. That was a study of the association of such diseases as leukemia with the amount of X-rays that have been received by people in their past medical histories.

I have forgotten the three states. I think Maryland was one of them, and two other states were represented, and some thirteen million people's histories that were culled for the evidence of leukemia and deaths, and that study was conducted among others by Gibson and by Bross.

Q. ***What was the Oxford Study?***

A. The Oxford Study was the study of Dr. Alice Stewart that we discussed earlier which showed the excess leukemia and cancer in children whose mothers had been irradiated just by the small amount of X-rays in a diagnostic exam during the pregnancy.

Q. ***And how many people formed the basis?***

A. Many millions is the basis.

I want you to know that all of the millions were not gone through in detail. Representative samples were there. It just wasn't necessary, but it has the basis of many millions of people.

BY MR. KACHINSKY: Okay, I would like to introduce—in my records it is Exhibit No. 8, the letter from the NRC to Dr. Gofman.

[***Marked and filed Exhibit No. 8 in evidence***]

BY MR. KACHINSKY: [*Continuing*]

Q. ***Okay, could you tell us what you have there as an exhibit?***

A. I have the letter addressed to me, September 11th, 1978, on stationery of the United

UNITED STATES
NUCLEAR REGULATORY COMMISSION
WASHINGTON, D. C. 20555

SEP 11 1978

Dr. John W. Gofman
Committee for Nuclear
Responsibility, Inc.
P. O. Box 11207
San Francisco, California 94101

Dear Dr. Gofman:

Thank you for sending us a copy of the article from the newspaper, The Day, from June 30. We were not aware of this article and appreciate your calling it to our attention.

The interview with us on which the article was based was some time ago. We have recently gone to the Commission with a staff paper on the topic of occupational exposure, and think you will find this staff paper a current and much more detailed picture of our views. A copy is enclosed.

Your letter questioned the context of the quotations in the article. The interview covered the broad scope of occupational exposure to ionizing radiation, only part of which is in facilities or activities licensed by NRC. There was particular emphasis on occupational exposures in the inspection, maintenance and modification of nuclear power plants and in the practice of nuclear medicine. Presumably because of limitations of space, and in recognition of the particular interests of local readers, the article deals mostly with exposure of workers at power reactors and thus applies the quotations in a somewhat narrower context than the interview. However, we feel that the authors have made a fair and generally successful effort to present the main thrust of what we said. Although much of the material in quotation marks is in fact a summary of a rather long discussion, we don't feel that what was said is misrepresented, recognizing that a writer must have some latitude in reducing a long interview to an article of reasonable length.

There are a few points in the article that we would like to comment on.

(1) The most serious flaw in the article is that it does not seem to properly catch the extent of our concern with the growth in total worker exposure, that is the collective dose, as compared to the exposure of individuals where the picture is generally better. Also, in discussing the growth in the collective dose, we are sure that we discussed this as being of concern not just because of genetic effects as the article says, but also somatic effects.

(2) The article comments on planned proposals to the Commission. You can refer to the staff paper for the specifics of what we proposed to the Commission. One point we discussed was the importance of an informed decision by radiation workers to accept exposure, and this is undoubtedly the basis for the somewhat paraphrased quote attributed to Goller. The evidence mounts that, within the range of exposure levels encountered by radiation workers, there is no threshold, i.e., a level which can be assumed as safe in an absolute sense. We have found in discussions with people both in the power industry and in the nuclear medicine field that many people in these fields honestly believe that the low levels of exposure permitted are without risk, which reflects that somehow the wrong message has been delivered, in spite of the fact that our regulatory program has been based on the prudent policy assumption that any amount of radiation has a finite probability of inducing a health effect, e.g., cancer. We brought out in the interview our concern that in the past the way the regulations were written and regulatory programs were established may be responsible for creating the impression among many workers that the levels of exposure permitted are completely without risk. We felt that it should be made clear to workers that there is some risk. The third explicit point in the article is just one way of doing that.

(3) We discussed with the reporter two epidemiological studies started by the AEC, one being the Health and Mortality Study, of which the Hanford study was part, and the other the Transuranium Registry. The article somehow seems to confuse and lump these together.

Dr. John W. Gofman - 3 - SEP 11 1978

(4) The quoted remarks about the problems of ingested or inhaled activity were not so much in the context of industry, but rather exposure of personnel during the atmospheric bomb testing program, a subject with which we believe you are much more familiar than we. As regards the regulated industry, our comment was to the effect that the very high level of uncertainty with respect to both "body burden" and neutron exposure had been recognized early and a number of effective measures taken to keep this type of exposure to extremely low levels. We did say that this area needed more attention and in particular needed to be taken into account in expanded epidemiological programs since much of the weapons test exposure has involved inhalation or ingestion. We have been told that the total number of people exposed in the weapons program over the years is quite large.

We could discuss a few other points, but think it would be basically just nit-picking. The article is fundamentally a good job of reporting on a complex subject and seems to us to reflect an effort to improve public understanding of some of the tough issues that have to be faced in dealing with nonthreshold pollutants; we think there is a growing awareness that radiation is only one of these. This certainly is an area which needs public attention and greater awareness of the difficult public health judgments that must be made in balancing the needs of society against the adverse impact of activities taken to meet those needs.

Sincerely,

Robert B. Minogue

Robert B. Minogue, Director
Office of Standards Development

Karl R. Goller

Karl R. Goller
Office of Standards Development

Enclosure:
SECY-78-415

cc: Mr. Lance Johnson
The New London Day
47 Eugene O'Neal Drive
New London, Connecticut 06320

States Regulatory Commission, Washington, D.C., such letter being signed by—apparently the third page of this letter is not here. It's missing.

The letter is signed by Robert Minogue, the Director of Standards Development of the Nuclear Regulatory Commission and Carl Goller, also of the Office of Standards Development.

It is a letter written to me in response to my query as to whether a newspaper interview recorded from them really reflected what they had said, where they said that there was no safe threshold of radiation.

And according to Dr. Minogue, it cited here, they said, "What we have found is by God there ain't no threshold. There are some die-hards who still believe in it, but it is a myth that there is a threshold."

And in the letter Dr. Minogue and Goller wrote me, they indicated that it says, "Although much of the material in quotation marks is, in fact, a summary of a rather long discussion, we don't feel that what it said is misrepresented, recognizing that a writer must add some latitude in reducing a long interview to an article of reasonable length."

And a little later they comment on the fact that they know of no evidence for a threshold, for a safe threshold.

So, that is the Nuclear Regulatory Commission stand indicating that they, too, agree with the rest of the scientific world that there is no evidence of a safe amount of radiation.

BY MR. KACHINSKY: I move to place that exhibit in evidence.

BY THE COURT: Exhibit No.?

BY THE CLERK: No. 8.

BY THE COURT: Exhibit No. 8.

BY MR. KACHINSKY: [*Continuing*]

Q. ***Could you explain briefly about the linear hypothesis and how that is used?***

A. Yes, when you irradiate a group of individuals with varying doses, starting with no extra dose added, no dose added, first we have some number of cancers that occur in people without any added radiation.

And suppose we add to people ten, twenty, thirty, forty, up to say a hundred units of radiation in the different groups. The number of cancers we know for sure goes up to higher numbers the more the dose.

If you plotted the dose on one axis and the number of cancers on the other, if it was a straight line, then you say that's a linear relationship. It just means the number of cancers is just directly proportional to the dose.

On the other hand, just by itself, you can't theorize there should be a straight line. It could rise faster than a straight line would indicate or slower.

But much of the evidence in the case that there is this straight line relationship with dose for leukemia in the people who are irradiated in Japan, for Alice Stewart's children, whose mothers were irradiated; for Mayes and Spees' study of children and adults who were irradiated with radium 224, straight line relationship; for many studies of breast tumors in animals induced by radiation, straight line relationships; more radiation is straight proportional to tumors.

And now Dr. Robert Kinnard and Dr. Hempleman

of Rochester have put together a variety of experiences on the induction of human thyroid cancer by radiation, including many children who were irradiated for certain other things and developed thyroid cancer, including the Marshallese who have developed an epidemic from thyroid cancer and radioiodine.

And putting together all these studies, they have found a straight line relationship between radioiodine dose and cancer. So it's direct proportionality.

I would like to point out it doesn't alter the fact of radiation causing cancer, whether it's a straight line or not. We don't depend on the linear hypothesis.

As a matter of fact, a fair amount of evidence is accumulating now that while people used to think you were being conservative and maybe overestimating the effect by the linear hypothesis, there is some evidence that is highly suggestive that we might have underestimated the effect by saying it is just a straight line.

Indeed, when I personally analyzed carefully the results on the uranium miners some eight years ago, and people had suggested that the linear hypothesis might overestimate, my data indicated, if anything, that at the lower doses it was worse than the linear hypothesis, and I published that report.

Q. ***Has the linear hypothesis been adopted by the Nuclear Regulatory Commision?***

A. The Nuclear Regulatory Commission and all other standard-setting bodies have generally said that they used the linear hypothesis.

They do not say it's proved to be the correct thing, but they do work with it as their way of predicting deaths.

Q. ***Okay, I would like to show you plaintiff's—I have it as Exhibit No. 9, "Re-Analysis of Data Relating to the Hanford Study of the Cancer Risk of Radiation Workers."***

BY THE CLERK: This will be Exhibit No. 9. For the record, Exhibit No. 9 is a study: "Re-Analysis of Data Relating to the Hanford Study of the Cancer Risks of Radiation Workers."

BY THE COURT: Let it be filed.

[*Marked and filed Exhibit No. 9 in evidence*]

A. Yes.

Q. ***Could you briefly explain—I know you have touched on this briefly earlier, but could you briefly refresh us on what this is?***

A. Well, this is the second phase of the analysis by Drs. Kneale, Stewart and Mancuso presented in March, '78 at Vienna of their analysis of the extra deaths; whether there are extra deaths in the workers at Hanford exposed to the so-called safe amount of radiation or not and how many extra deaths.

Between their first report given at the Health Physics Society and this one, they had analyzed additional cases, in their study, which I think we don't want to go into the enormous number of tables and so forth, but the fundamental conclusive finding of this study, and one that I have tested independently is the fact that for every thirty-three units, rems, of total body radiation, you will add a number of cancer deaths equal to the spontaneous number. We call that sort of number a doubling dose.

This is a study which I have done independently, not using any of their analysis at all but by my own method, and I arrived at thirty-eight, which, as I said earlier, is good agreement.

Q. ***Are you familiar with Dr. Mancuso?***

A. Yes.

BY THE COURT: Dr. who?

BY MR KACHINSKY: I asked the witness if he's familiar with—

BY THE COURT: Dr.?

BY MR. KACHINSKY: —Mancuso.

BY THE COURT: Okay.

A. He's one of the three authors of this study.

BY MR. KACHINSKY: [*Continuing*]

Q. ***What has happened to him since this study?***

A. He had his funds taken away by the Department of Energy.

There was a Congressional hearing about that, and the reasons for taking them away were referred to by Representative Paul Rogers as the weirdest type of reasons he had ever seen.

Q. ***Has Congress been investigating these, the figures of this report?***

A. Congress been investigating these figures?

Congress doesn't, I think, really have an arm that investigates scientific matters of that sort, but Congress has expressed its serious concern and may recommend that some government agency launch a very much larger study of all atomic workers to get even a bigger base of evidence and to get the matter out of the Department of Energy's hand in view of the Department of Energy's one-sided and apparently biased view on radiation injury.

Q. ***Are you acquainted with Dr. Samuel Milham?***

A. I do not know Dr. Milham. I have heard about him and his work.

Q. ***What does his work involve?***

A. He's a public health statistician in Washington, and he continually looks at cancer figures in the State of Washington, and before the Mancuso study was ready, Milham indicated from his studies that there was too much cancer occurring in the workers in that part of Washington where Hanford is; that is not a study of the relationship of radiation to cancer, just that there seem to be too many.

Q. ***As a result of your education and research in relevant areas, would you say that the nuclear fuel cycle would ever contain the radio-nuclides and isolate them from the biosphere?***

A. As a result of all my studies and my scientific experience in laboratory and semi-engineering and engineering projects with such things, I find it not credible that these materials can be contained perfectly, not at all credible to me.

Q. ***And what would be the effects of the failure to isolate these materials from the biosphere?***

BY THE COURT: He has already said that two or three times—

BY MR. KACHINSKY: Okay.

BY THE COURT: —up one side and down the other.

BY MR. KACHINSKY: [*Continuing*]

Q. ***Would you say that there has been conclusive evidence to the effect that low-level ionizing radiation, such as that produced by the nuclear fuel cycle, has been, in fact, shown to cause cancer?***

A. I feel the evidence is conclusive that low-level amounts of radiation, even below the dose allowed, does cause cancer, yes.

Q. ***Would you like to say the public has been adequately or correctly informed about the consequences to health of the nuclear fuel cycle?***

A. The public has been pretty badly deceived about the consequences of a nuclear fuel cycle, and we urgently need an honest statement to the public.

Q. ***You have testified before the NRC, the Nuclear Regulatory Commission?***

A. I have not testified before the NRC.

I have prepared documents for the Sierra Club and the National Resources Defense Council in connection with certain matters they had before the NRC, and my reports were put in the NRC docket on those matters, and I have prepared a statement for the GESMO hearings we referred to earlier on behalf of the Public Interest Group of Washington; that also was before the NRC before the President's action cancelled that hearing.

But I have not personally appeared at an NRC hearing.

Q. ***Do you feel their proceedings reflect standard scientific procedures?***

A. I have a strong objection to NRC proceedings, and that's one of the reasons I won't participate in them, because I do not believe they really address the crucial scientific issues.

For example, in some of the hearings the key issues concerning safety and health and cancer are excluded automatically from the hearings, and I think the hearings become a sort of an absurd sham in view of the fact that there is a limitation of the subjects that can be taken up concerning a specific nuclear power plant. They just ask how does this nuclear power plant meet or not meet the specifications when the citizens of that region are concerned about the health implications, and whether the standards are right and whether the evidence on low-dose cancer production is right, and those things are always excluded from such hearings.

BY THE COURT: How were the Commission's standards set, Doctor?

A. Sir?

BY THE COURT: What was the procedure by which the Regulatory Commission set its standards or arrived at them?

A. They arrived at their standards from the recommendations of bodies of scientists that have gotten together and looked at all of the scientific evidence from around the world on the production of human cancer and leukemia and tried to figure out how many cancers per unit of radiation, and the NRC has pretty much followed the suggestions that came from such bodies as the International Commission on Radiological Protection and the National Council on Radiation Protection.

In essence, for example, in following the International Commission's recommendations, the International Commission never said there was a safe amount of radiation. They recommended certain guidelines that they thought were consistent with giving the developing atomic industry latitude, and the Nuclear Regulatory Commission has followed the guidance of the International Commission and any scientific evidence that came up since any such recommendations, and I know that they continue to look at the new scientific evidence in considering whether they wish to alter their standards.

For example, there is before them right now a petition to reduce the occupational dose, and the NRC staff is doing great and extensive study to see whether that ought to be granted or not.

BY THE COURT: Well, is the purpose of this lawsuit to have this Court determine the standards are inadequate?

A. It is my understanding that the purpose of this lawsuit is to demonstrate that the standards in force give permission to give doses that are killing people and will kill people and to raise the question of the constitutionality of those standards that permit the killing of people.

BY THE COURT: The Commission has set standards with which you disagree and with which a great number of people disagree, and maybe with which I disagree as far as that is concerned, but they have, after having made an examination, heard opinions, reports, had hearings and what-have-you, set certain standards which they feel are reasonable under the circumstances, is that correct?

A. Yes, and, for example, in the table I read from the current nuclear industry, they said they expect within the next twenty-five years eleven hundred people will die of cancer.

Now, my disagreement is that I think the number is higher, but what they have said is that eleven hundred is the number who would die by their standards.

BY THE COURT: A lot of people get killed on the highways when they drive automobiles, don't they, at rates of speed?

A. Yes, they do, sir.

BY THE COURT: And the higher the speed, the more people are killed?

A. That is correct, sir.

BY THE COURT: And does that make then the standards set by the Legislature of the State of Tennessee unconstitutional? Is there something similar there or am I missing something?

A. I think there is something different in one sense.

BY THE COURT: The thing that bothers me about this while procedure is that—and, believe me, I know nothing about radiation. I'm not supposed to know anything about it, except what I read in the newspaper, and that's—but anyhow, health and welfare of the people on the highway and health and welfare of people as a result of energy requirements—is there something there I'm missing somewhere?

A. Well, there is one thing, Your Honor. In the case of an automobile—and I would be the last one to condone the speed limits that have caused such numbers of deaths—individuals drive on the highway by their own choice, but children and other people who are in their homes and get irradiated from the nuclear fuel cycles and die fifteen to thirty years prematurely, did it without any vote, decision or anything else, as the driver on the highway does . . .

BY THE COURT: You have got a point there, except an awful lot of cars run off the highway and hit houses.

A. Yes, that is true, and that I think is a moral and ethical question that deserves the most careful examination in our society.

But in the case of—I know of no government agency that licenses murder, and if you license a plant to emit these things, you know the murders are going to occur. There is always a chance with a car . . .

BY THE COURT: How about the death penalty?

A. That is constitutional, in my understanding, on the ground that due process of law has been exercised, but in this case there has been no due process.

BY THE COURT: Okay, go ahead.

BY MR. KACHINSKY: [*Continuing*]

Q. ***Would you say, considering what you know about the nature of radioactive particles and emissions, that there is a reasonable probability that the plaintiff, Jeannine Honicker, will be struck by particles emitted as a result of some phase of the nuclear fuel cycle?***

A. Yes.

Q. ***And the possible results—what would be the possible results of being struck by such a particle?***

A. She has an increased risk, risk of cancer or leukemia as a result of this; that doesn't mean she's going to get it, but she has been damaged, and every day that she gets struck again she has been damaged more, and her risk of cancer or leukemia that would otherwise not occur does get enhanced.

Q. ***Okay, I would like to show you—I have marked as Exhibit 10. It is the Annual Report to Congress of the Energy Information Administration.***

BY THE CLERK: For the record, Exhibit No. 10 is an annual report study.

[***Marked and filed Exhibit No. 10 in evidence***]

PROJECTED HEALTH EFFECTS OF SELECTED RADIONUCLIDES OVER TIME.

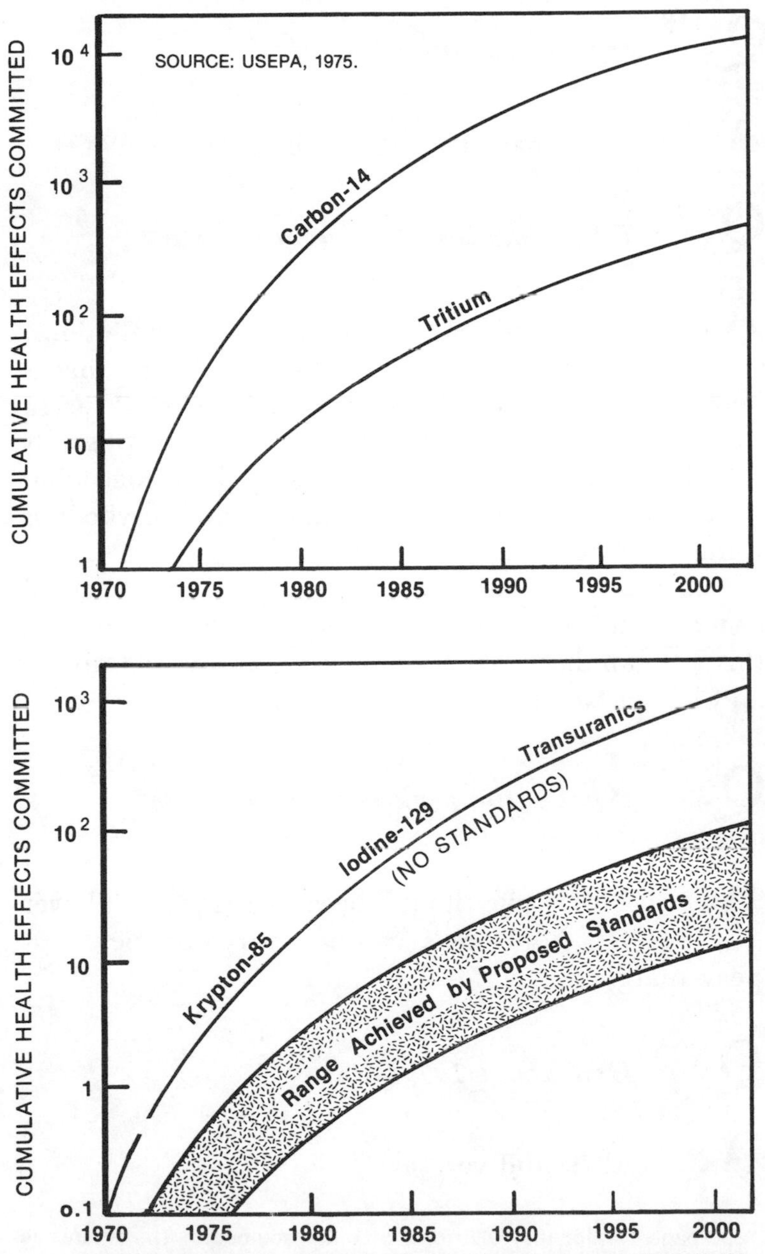

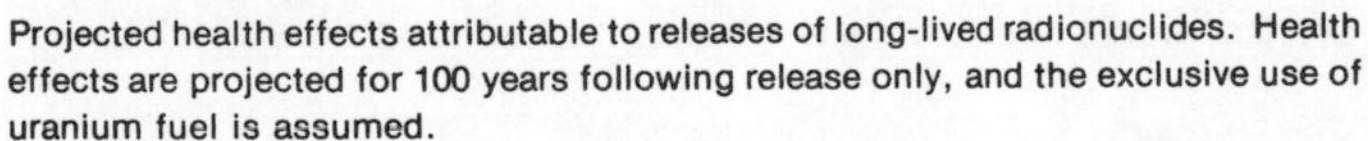
Projected health effects attributable to releases of long-lived radionuclides. Health effects are projected for 100 years following release only, and the exclusive use of uranium fuel is assumed.

BY MR. KACHINSKY: [*Continuing*]

Q. ***Are you familiar with this?***

A. Yes, I have looked at this study, yes.

Q. ***What does this document refer to?***

A. It refers to the amount of energy installations, power installations by the type of fuel that is being used and an estimate of just how much extra installation we have over what we need; that's usually described as the margin above the peak load that occurs, say, at the height of a time when everybody is using their power to the fullest.

You need to have—to have a secure electrical system, you need to have some margin above that, and this is a tabulation of such margins from 1966 to '76 and forecasts out to 1995.*

Q. ***What is the peak margin for 1978?***

A. This is an estimate, because that's not all over yet, 30.8—let's call it thirty-one per cent peak margin.

Q. ***How about 1979?***

A. 1979, did you say?

*See page 177 for the 1978 form of this document. The figures in Dr. Gofman's testimony are from the 1977 tabulation.

Q. *Yes, 1979?*

A. 28.6 per cent.

Q. *1985?*

A. Twenty—19.6 per cent.

Q. *And 1990?*

A. 18.1 per cent.

Q. *What per cent of the present generating capacity is provided by nuclear power?*

A. Approximately twelve or thirteen per cent, I think, if all of the nuclear plants are working, our present generating capacity will be twelve or thirteen per cent.

Q. *What does this mean in the event that nuclear power is dropped from the capacity of this country?*

BY MR. EILPERIN: Objection, Your Honor. As far as I know, Mr. Gofman has not been established as an expert in the electrical power industry and production.

BY THE COURT: You are probably right. I will let him give his opinion though.

A. It is my opinion it would not be at all serious, that with moving electric power around and the various things that we have available and the shifting from use of electricity for heat very foolishly into using other ways and conservation, that we would suffer hardly at all from turning off the nuclear power plants and not building any more.

Of course, that depends on using interties and moving power around, and there might be a brief period of having to make some arrangements, but I believe with this margin of excess capacity and the use of moving power around, that we can certainly cope without any dire effects if we eliminate nuclear power.

BY THE COURT: Twelve per cent cut in the total . . .

A. In the electrical, not in the energy; two and a half to three per cent in the energy of the country.

BY THE COURT: But ten per cent—twelve per cent in the electrical production of this country would not cause a problem?

A. Because we have such a large excess capacity. You see, that's what we have here is a thirty-one per cent excess capacity right now.

BY THE COURT: Mr. Witness, you may believe that. I don't.

A. I think that's a subject for debate between many people.

BY MR. KACHINSKY: [*Continuing*]

Q. ***Okay. Have you studied the economic situation with respect to nuclear power?***

A. I have done and published studies on the electrical power output per short ton of uranium oxide and done an analysis of all of the factors of loss of uranium oxide in the nuclear fuel cycle, which allowed me to come up with some numbers as to how much electrical energy you could get out of a given amount of uranium. Those studies also led me to estimates of the available amounts of uranium to generate power by the current cycle.

And as a result of knowing really how much electricity you could get out at the most, and comparing them with the studies of how much has actually been obtained from those plants which have been operating, I could help make estimates concerning some of the economic factors of electrical power.

Q. ***Well, what are your conclusions?***

A. My conclusions are that we have a very serious problem of finding enough uranium to fuel the currently projected reactors, and I think that's being manifested in the rising cost of uranium, such that the economics of nuclear power have continued to look more poor with the passage of time.

I have studied the capital...

BY THE COURT: If that's so, will that not just eliminate the problem, because if we can't afford them, we will quit using them?

A. I think that is a very separate problem, and the marketplace is speaking very strongly to that issue.

As you know, Your Honor, the electric utility industry has become increasingly gun-shy of ordering nuclear plants, and they have been cancelling plants at a great rate, and I believe the economic marketplace will kill nuclear power, unless a subsidy is produced by the government. There have been so many subsidies to the industry already that it has been running on tax dollars in effect.

But I don't think that is the issue we are discussing here today. It could cure itself, but the issue is really whether the people—whether someone can be licensed to allow—to permit death.

BY MR. KACHINSKY: [*Continuing*]

Q. ***Are the future consequences of the nuclear fuel cycle predictable with any great degree of accuracy at this time?***

A. I think I have answered that before. I've said that I cannot predict whether one per cent, one-tenth of a per cent, a thousandth of a per cent, or a millionth of a per cent of the radioactivity will get out; that's just a gambling game.

The more that gets out of control into the biosphere, the more deaths, and I cannot, therefore, predict the consequences with any accuracy, except to say that there are going to be deaths. There are deaths occurring now, and there will be more, the more that gets out of containment.

Q. ***Thank you. I have no more questions.***

CROSS-EXAMINATION

BY THE COURT: Cross-examine.

BY MR. SLAGGIE: Your Honor, with your permission, I think I have a loud enough voice that can be heard. I would prefer to cross-examine the witness from here.

BY THE COURT: I want it on that gadget there, because we tape everything here.

BY MR. SLAGGIE:

Q. ***In that case speak up so I'll be sure to hear you.***

BY THE COURT: In addition to what the Court Reporter takes down, we also put it on a tape, so if I want to listen to it later on, he doesn't have to come into my office and sit there and read it to me. All right?

BY MR. SLAGGIE: Yes, sir.

BY THE COURT: Okay.

BY MR. SLAGGIE: [Continuing]

Q. ***Mr. Gofman, I notice from your statement of qualifications that you have done independent research in nuclear physical chemistry.***

A. That is correct.

Q. ***Have you also done independent research in medicine?***

A. Yes. I have devoted most of my life to independent research in medicine.

I have written three books on heart disease and published a hundred thirty-seven scientific papers on heart disease and arteriosclerosis, and I have received the two highest American awards for my medical research on heart disease.

Q. ***Is this experimental work you have done yourself?***

A. Yes.

Q. ***At the beginning of your testimony, you described what I believe was a theory or mechanism by which radiation can cause cancer of human beings, is that correct?***

A. What I said was that the fact is that radiation causes cancer in human beings and that we speculate on a mechanism, being radiation affecting the genes and chromosomes.

I described it as a possible explanation, and none of the statements that radiation causes cancer hinges on that mechanism or theory.

Q. ***Well, you have said 'a speculation.' Is this a mechanism from which we could now calculate, actually predict what the risk of cancer would be from radiation?***

A. No, we calculatc thc risk of cancer from radiation from dead bodies caused by radiation.

We have the evidence now. We don't use that mechanism to calculate.

Q. ***Put less colorfully, do you mean that we calculate the risk of radiation based on statistical data rather than a basic theoretical understanding?***

A. That is correct, yes.

Q. ***You mentioned a number of times—well, I think you said that the evidence showed beyond a doubt—***

A. Yes.

Q. ***—that if we consider two groups, one of which has been irradiated, the other has not been, that the irradiated group will catch cancer, is that correct?***

A. That is correct.

Q. ***And I believe you cited the data by Stewart.***

A. Stewart, that's one of them.

Q. ***Low level doses causing cancer, is that correct?***

A. That is correct, the doses associated with diagnostic radiation, which were in the range of three hundred millirems to one and a half rems to the fetus.

Would you say again what the range of the radiation was?

A. Three hundred millirems; that's approximately what it was for one film—up to one and a half rems, which is what it is for five films delivered to the mother, and that is the range in which her studies were done.

Q. ***Over what period of time was that radiation delivered?***

A. It was delivered in either one, two, three, four, or five X-ray films, and an X-ray film is a fraction of a second, all done during the third trimester of pregnancy.

Q. ***Let me be sure I understand that. You are saying the data, the Stewart data involves doses of something like three hundred millirems—that's what, three-tenths of a rem—***

A. Yes.

Q. ***—delivered to a fetus in a fraction of a second?***

A. Yes, and Dr. Stewart's other studies are the study of the Hanford data, where that same amount of radiation is delivered over many, many months in the course of their work.

Q. ***That's what I wanted to ask you. Let me try to give you a hypothetical, if I might, to understand this.***

Suppose hypothetically we are talking about

some activity that would give every individual in a large population a dose of no more than three-thousandths of a rem—three millirems, I believe that's right,—

A. Yes.

Q. ***—over a year's period.***

A. Three millirem.

Q. ***All right, three millirem over a year's period—***

A. Yes.

Q. ***—to a large population.***

A. Yes.

Q. ***Would you say—would you extrapolate the Stewart data to conclude, using your words, that beyond a doubt this smaller radiation dose spread over a long period of time would cause cancer?***

A. I made my statement quite explicit. I said . . .

BY THE COURT: Either yes or no. Come on, don't—either say yes or no.

A. I would say yes, that I—but I can't say it's beyond a reasonable doubt. Some people doubt that.

Q. ***Dr. Gofman, you mentioned at one point, you referred to the dangers of plutonium.***

A. Yes.

Q. ***And plutonium released from reactors.***

A. From the nuclear fuel cycle.

Q. ***You cited, I believe, the example of the Rocky Flats plant as an example, where a large amount of plutonium, relatively large amount of plutonium had been released. Maybe I didn't hear you, but did you specify whether or not Rocky Flats was a weapons facility?***

A. I did not specify, but I would be able to specify that it is, and they handle plutonium; that's the same plutonium that would be handled in the fuel cycle.

That's not then related to the nuclear fuel cycle?

A. Rocky Flats isn't, but the plutonium is exactly the same kind of plutonium.

Q. ***The plutonium is exactly the same, but the release you cited was not a nuclear fuel cycle release, is that correct?***

A. Yes.

Q. ***Are you aware personally of any nuclear fuel cycle releases of plutonium likely to take place in the next two months that are comparable to these Rocky Flats releases you have cited?***

A. I'm not aware of any that are going to occur in the next two months.

However, there are facilities handling plutonium oxide in preparation for experimental assemblies of mixed oxide fuel.

For example, at Vallecitos, California, there are as much as five kilograms of plutonium oxide in house, and a release could be vastly more than occurred at Rocky Flats if released.

Q. ***You are saying this plutonium oxide is going to be used for fuel for the nuclear fuel cycle in the near future?***

A. It is being prepared for fuel rods that are part of the breeder program, yes.

Q. ***Is it your impression that the Commission presently permits the use of mixed oxide fuel?***

A. It is not only my impression—I know that General Electric is handling that plutonium for the preparation of fuel rods for experimental purposes right at the present time, yes, sir.

Q. ***But you are not contending that that plutonium in the near future is about to go into one of the current light water reactors?***

A. I most certainly am not.

Q. ***Dr. Gofman, I will show you an exhibit which I believe you testified from.***

A. That is correct.

Q. ***This is the chart table—what is the table number?***

A. S-5.

Q. ***Right, S-5. You refer to this as, please, Nuclear Regulatory Commission Data, which you called—correct me if I am wrong—The Expected Deaths from the Nuclear Fuel Cycle, is that what you called it?***

A. It said Estimated Health Effects from the U.S. Light Water—

Q. ***Right.***

A. —estimated.

Q. ***Did you refer to that as expected deaths?***

A. I would refer to it as an estimate as expected, if the cycle goes on, yes.

Q. ***All right, but those are estimated health effects, is that correct?***

A. Yes, yes.

Q. ***And do you happen to know what model those—you talked about the linear hypothesis.***

A. Um-hm.

Q. ***Are these estimates, to your knowledge, based on the linear hypothesis.***

A. To my knowledge they would be, yes.

Q. ***They are based on the linear hypothesis?***

A. But the linear hypothesis is not a singular thing. It depends upon what increments in number of deaths per rem you use to plug into the linear hypothesis that determines the numbers.

For example, I, too, use the linear hypothesis, and I would get larger numbers than these.

The linear—a straight line can be this steep or much less steep or much more steep but still be the linear hypothesis.

Q. ***Would it be fair to say that these data are consistent with the hypothesis of radiation of biological effects advanced by the BEIR Committee in the 1972 report?***

A. It would be fair to say that.

Q. ***Did the BEIR Committee also observe—I believe also somewhere in their report—that the data could not exclude zero effect at very low doses?***

A. They have studied that.

Q. ***I see. Are you familiar with the Honicker petition—***

A. Yes, sir.

Q. ***—which is the subject of the complaint today?***

I noticed you mentioned—referred to data by Thomas Mancuso and by Irwin D. Bross—

A. I did.

Q. *—which, I believe, are cited in the Honicker petition, is that correct?*

A. And I referred to my own analysis.

Q. *Are you also familiar with the data by Ernest J. Sternglass?*

A. Some.

Q. *Do you believe that the data by Ernest J. Sternglass is reliable, and would you also place faith in that as you have in the Mancuso . . .*

A. Ernest Sternglass has published enormous numbers of reports over the past seven years, and you would have to specify which one you want me to answer to.

Q. *All right, in the Honicker petition—if I can find it. I will have to look, but I can ask it now. There is data referred to in the Honicker petition with regard to releases from the Millstone plant.*

A. Yes.

Q. ***Are you familiar with that?***

A. Yes.

Q. ***Do you regard that as reliable?***

A. I have no reason to regard it as unreliable. I am still in the act of studying that, but I certainly have no reason to regard that as unreliable. I have not seen anyone challenge the actual numbers of the measurements of the numbers of curies of strontium 90, and I am very interested in the statistics on cancer effect there.

With such matters, there is a way to find out. One watches the cancer rates by year, and I would like to see more data for this year and next year and such, and I have no reason to reject that study at all.

Q. ***And at this point then you have no opinion one way or another?***

A. No, I do not, and I have no reason to reject it. Sometimes when I make a study, I can reject it.

Q. ***I'm sorry, sir?***

A. I said sometimes I can look at data and say I do not believe this, but I have no reason to reject that study at all.

Q. ***No reason to . . .***

A. To reject it.

Q. ***To reject the study. Are you familiar with what the NRC has said with regard to that study? Are you familiar with any comments the NRC has made about it?***

A. I'm not very familiar. I think I have seen some comments, but I don't know them.

Q. ***I believe you were asked whether the standards, NRC standards now in effect would permit some people to die as a result of the nuclear fuel cycle. Were you asked that?***

A. I believe I was asked something . . .

Q. ***And what was your answer to that?***

A. I believe the standards now in effect, the permitted doses would permit people to die, yes.

Q. ***What standards, in your view, would not permit people to die with regard to releases from the nuclear fuel cycle?***

A. Zero release.

Q. ***Zero radiation. And that is the only standard . . .***

A. That would keep people from dying, yes, sir.

Q. ***In your view, is there anything, any activity, that can justify releasing radiation to the general environment where people may be exposed to it?***

A. I think that that is a very fundamental question for society.

The Constitution tells me that it's not permitted to do something that takes life away without due process, and releasing radiation does that.

So, I can't justify a way if it justifies releasing radiation and killing people unless we change the Constitution.

Q. ***Is it your testimony that no dose however small, as long as it is not zero—the possibility, rather, of injury and death for any dose, however small, but non-zero, cannot be excluded, is that correct?***

A. The answer is, yes, that cannot be excluded.

Q. ***Well, aren't there a number of activities the government engages in—well, private industry—I would say activities generally engaged in that do result in radiation doses to individuals?***

A. Yes. There are.

Q. ***—in addition to the nuclear fuel cycle?***

A. Yes, there are. There are medical uses.

Q. ***Doesn't medical—there are medical radiation doses which are direct exposures. Would you regard those as justified, assuming they are properly administered?***

A. I have written extensively on that subject.

BY THE COURT: Well, yes or no? Are they justified under certain circumstances?

A. Yes.

BY MR. SLAGGIE: [*Continuing*]

Q. ***They are justified?***

A. Under certain circumstances, yes.

Q. ***Well, as a physician—are you familiar with some of the techniques using radioisotopes that have been developed in medical diagnosis and treatment over the years?***

A. Yes, sir, I have worked in that.

Q. ***In your view are those valid medical techniques?***

A. I'm not sure at the present moment whether we are killing more people than we are helping with them. In that case I would consider them invalid.

Q. ***Can you—would you—let me ask you how are these radioisotopes produced?***

A. Some of them are produced in cyclotrons, for example, because there's no convenient way to make them by neutron activation. Some of them can be made—some of them are fission products of the nuclear fuel cycle.

And, lastly, some are what we call neutron activation products, where you have the neutrons in a reactor. You put in a parent material, and you can create the radioisotope by the irradiation and then they're isolated at a place like Oak Ridge and distributed for use in medicine.

Q. ***This reactor that produces the neutrons for neutron activation, does it have some of the same problems of the power reactors we have been***

talking about with regard to releases to the environment?

A. Oh yes, all reactors have the possibility of releasing things.

Of course, a thousand megawatt reactor is a very large reactor in terms of the amount of radioactivity it produces.

Q. ***But a small research reactor nonetheless does emit small doses?***

A. They do.

Q. ***Do you regard that as defensible?***

A. I think one has to consider the seriousness.

Q. ***What?***

A. I think one has to consider the seriousness of those.

Q. ***How about the cyclotron? Doesn't operation of a cyclotron activate some?***

A. Yes. I have personally received radiation from cyclotron radiation. I'm not lethal by the irradiation I received thereby.

Q. ***How about persons who are associated with the cyclotron, people who are just in the area, can't they nonetheless receive some small doses?***

A. They can, because some products are made radioactive, and those products can come out from the cyclotron.

Q. ***The High Energy Physics Research Program, for example, uses very large particle accelerators. Wouldn't they consider a certain amount of those—***

A. Yes.

Q. ***—to the air?***

A. To the air—well, that's getting very, very small. The workers do get some dose.

Q. ***With these very small quantities, nonetheless, you cannot exclude the possibility that they will . . .***

A. Oh, no, I cannot exclude that possibility.

Q. ***I think you have made—well, basically returning both to your affidavit and to a remark you made on the stand. I think you applied the word 'murder' to those who would intentionally***

release radioactivity to the atmosphere or to the environment.

Would you be willing to say, "Now, that seems a bit strong?"

A. That's still the word I would use.

BY MR. SLAGGIE: I have no further questions.

BY THE COURT: All right. Call your next witness.

BY MR. KACHINSKY: Can I ask the witness on redirect?

BY THE COURT: Yes, if it's something he covered, nothing that you forgot.

REDIRECT EXAMINATION

BY MR. KACHINSKY:

Q. ***Did your co-workers at Livermore Laboratory examine some of Dr. Sternglass' data?***

A. At which laboratory?

Q. ***The Livermore Laboratory?***

A. Oh, there is one Sternglass study that was put out in the '60s concerning the causation of infant mortality by fallout from weapons' tests. That study was sent to me as Associate Director of Biology and Medicine at the Livermore Laboratory, sponsored by the AEC, and I was asked if any of our people could comment on it.

Dr. Arthur Tamplin of my staff did comment on it and wrote an indication that he thought Sternglass had overestimated the number of deaths caused by fallout and indicated why, and it is of great interest to this Court that the data that have become available since, and that are in the Petition, demonstrate that it is Dr. Sternglass' analysis that seems more correct than my colleague, Dr. Tamplin's.

Q. *Thank you. I have no further questions.*

BY THE COURT: Step down. Call your next witness.

BY MR. KACHINSKY: I would like to call Dr. Ernest Sternglass to the stand.

BY THE COURT: Now, are you going to go through the same thing with him?

MR. KACHINSKY: No.

BY THE COURT: What are you going to ask him?

BY MR. KACHINSKY: I'm going to ask him specifically about studies that he did about the nuclear radiation cycle and on the nuclear plants in Connecticut.

BY THE COURT: Well, is he going to say basically what the other fellow said?

BY MR. KACHINSKY: I believe he has additional testimony to give.

BY THE COURT: You will not ask him any questions you asked the other fellow, because I am not going to have just cumulative testimony. Ask him something different.

* * * * *

DR. ERNEST J. STERNGLASS, having first been duly sworn, was thereupon called as a witness and testified as follows, to wit:

BY MR. KACHINSKY: I have a biography of Dr. Sternglass that I would like to enter into evidence.

BY THE COURT: Exhibit No. 11, I think, already filed.

BY THE CLERK: For the record, Exhibit No. 11, biography of Dr. Ernest Sternglass.

[***Marked and filed Exhibit No. 11 in evidence***]

Biography of Ernest J. Sternglass, Ph.D.

Educational Background

Bachelor of Electrical Engineering, Cornell University, 1944
Master of Science in Engineering Physics, Cornell University, 1951
Doctor of Philosophy in Engineering Physics, Cornell University, 1953

Occupational Background

Professor and Director of Radiological Physics, University of Pittsburgh, 1967-present
Advisory Physicist, Westinghouse Research Laboratories, 1952-1967
Research Physicist, U.S. Naval Ordnance Laboratory, White Oak, Maryland, 1946-1949
Electronics and Radar Technician, U.S. Navy, 1945-1946

Professional Societies and Awards

Elected to Eta Kappa Nu (electrical engineering honor society)
Sigma Xi (scientific research honor society)
McMullen Fellowship, Cornell University
Fellow, American Physical Society
American Association of Physicists in Medicine
Radiological Society of North America
Society of Nuclear Medicine
Health Physics Society
Federation of American Scientists
American Association for the Advancement of Science
Philosophy of Science Association
American Astronomical Society
Past Chairman, Pittsburgh Chapter of the Federation of American Scientists

Professional Experience

- Development of nuclear and radiological instrumentation for diagnostic purposes in nuclear medicine and radiology
- Fundamental theory development on the interaction of radiation with matter and its biological action
- Statistical studies of the effects of low-level environmental radiation on the human fetus, infant, and adult
- Author, *Low-Level Radiation,* Ballantine Books, N.Y., 1972, translated into German and Japanese
- Author of a review article on the biological effects of environmental radiation from natural and man-made sources (Chapter 15), *Environmental Chemistry,* J.O'M. Bockris (ed.), Plenum Press, N.Y., 1977
- Author of some 100 scientific papers and articles in the field of nuclear physics, radiological sciences, nuclear instrumentation, astronomical instrumentation, radiation interaction with matter, and biological effects of radiation on man

Patents

Ten patents in the field of electronic and nuclear instrumentation

DIRECT EXAMINATION

BY MR. KACHINSKY:

Q. ***Have you published articles, books, and scientific papers relevant to radiation and its effect on health?***

A. Yes, I have.

Q. ***Is this book "Low Level Radiation" a book you have written?***

A. Yes, it is.

Q. ***When did you write this book?***

A. In 1969 to '71.

BY MR. KACHINSKY: I would also like to ask Dr. Sternglass if he feels in the history of radiation protection, if he feels the present standards are adequate and why.

BY THE COURT: I'm sure he will say no, because you have got him on the stand, but you can ask him.

BY MR. KACHINSKY: [*Continuing*]

Q. ***Would you answer that question, Dr. Sternglass?***

A. I would say, Your Honor, that is correct.

BY THE COURT: I don't mean to infer that you would change your testimony just because you are on the stand, but I mean to say he wouldn't waste his time putting somebody up here that was going to testify to the contrary, is what I meant to say. Do you understand why...

A. Yes, sir.

BY THE COURT: All right.

A. I mean to add that, of course, the grounds or the reasons for my belief are different than Dr. Gofman's. They are based on quite different kinds of studies, and I would like to be able to explain those reasons.

BY THE COURT: Have at it.

BY MR. KACHINSKY: Okay.

BY THE COURT: Just go ahead. Don't wait for a question. Just tell us about it.

A. The basic reasons why I have now reluctantly over the years come to the conclusion that our standards for radiation in the environment are not adequately low arises from a period of years of research and study which I have carried out in which I examined actual statistical data on the increases in infant mortality and the leukemia and cancer rates following the detonation of nuclear weapons during the 1950's, beginning in 1945 at Alamogordo and continuing to this day in decreasing amounts.

My studies simply try to relate the changes that were observed in the number of babies that died, the numbers of babies that were born underweight, the numbers of babies that were born congenitally defective and mentally retarded during the period of heaviest nuclear testing.

I became involved in this in the early '60s when I was asked by members of the Federation of American Scientists in the Pittsburgh area to examine the consequences of nuclear fallout, whose radiation levels are comparable to those from the environment, typically less than a hundred millirems per year.

At the time I published a paper in the *Journal of Science* in June of 1963, in which I indicated that based on the earliest studies of Dr. Alice Stewart and Dr. Kneale of Oxford University, one would expect to see a significant increase in leukemia and childhood cancers all over the world from the calculated doses from bomb test fallout.

Thereupon I was asked by Congress to testify on the subject in August of 1963 at hearings held by the Joint Committee on Atomic Energy at which time I was asked to examine and explain the serious potential implications. We were dealing here with very small amounts of radiation given over long periods of time, which, of course, is precisely the kind of situation that we encounter in the peaceful nuclear cycle, where again we have very small amounts of

CANCER MORTALITY RATE FOR 5-9 YR. OLD MALES IN ALL OF JAPAN

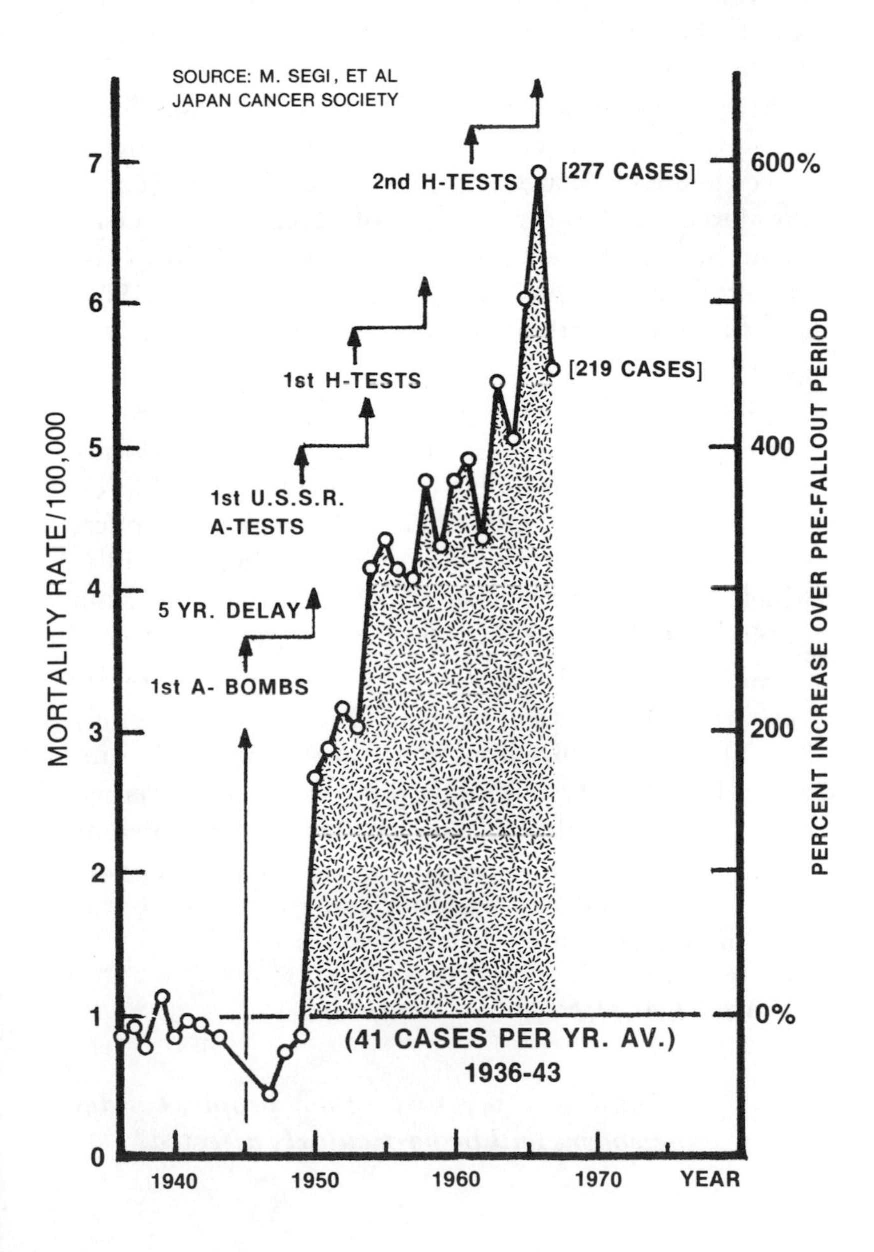

radiation, typically ten, twenty millirads, ten, twenty per cent of background per year. Until that time it was widely believed that there would be no additional cancer deaths.

My suggestion was at the time that there was likely going to be an increase in cancer rates, and, indeed, since then data that was published by Dr. M. Segi of the Japanese Cancer Society of Tohoka University Medical School, showed a very sudden and sharp rise in childhood cancers in Japan between 1945 and 1952, and this was something on the order of a few hundred per cent.

The cancer rate came down again after the end of testing, and I'm glad to say that all over the world my researches have shown that from these very small levels of radiation previously believed to be harmless cancers are now beginning to come down rapidly, which leads me to believe that my initial prediction published in June of 1963 is probably correct.

And since these levels were well below those presently permitted from the nuclear fuel cycle, namely five hundred millirem max to any individual, and a hundred seventy millirems per year to the average person in the population, I now believe that the present permissible limits will produce a significant increase in the number of deaths among babies, the newborn, and, of course, cancers at all ages.

BY MR. KACHINSKY: [*Continuing*]

Q. ***Would you say any other people besides newborns will be particularly affected?***

A. There are now studies which I have begun to carry out and which are quite separate from those of Dr. Mancuso and Dr. Stewart that show that older individuals, over age fifty-five or sixty, are also much more susceptible to the development of cancers and leukemia. This came as somewhat of a surprise to many people in the field, but it now appears that there are some individuals in our society, as has been pointed out in other studies of Dr. Bross published in the *New England Journal of Medicine*, who happen to be particularly susceptible. People who have had a history of allergic or infectious diseases, especially among children, have been clearly shown to have sometimes a five, tenfold greater risk than normal of developing cancer or leukemia.

BY THE COURT: Are there people in the medical field that disagree with you?

A. Sir, there is not a scientific question in which there is not considerable disagreement.

BY THE COURT: So for everyone that is on your side, there's somebody else on the other side, is that right?

A. Not exactly. Today, I believe, the overwhelming majority of independent non-industry scientists would say that low-level radiations probably lead to an increase in genetic defects, early difficulties with childhood problems, diseases in early childhood, leukemia and cancer, and I think that is, I would say, accepted by the overwhelming majority of the world medical profession.

BY THE COURT: Do the standards set by the Commission...

A. The...

BY THE COURT: Wait a minute. Do the standards set by the Commission recognize that fact?

A. I believe that they do not adequately recognize the latest findings.

BY THE COURT: But do they recognize the fact regardless of the latest findings?

A. Oh, yes.

BY THE COURT: They recognize the fact that radiation will cause it?

A. Yes, they do.

BY THE COURT: Then the difference then is the difference between you and the standards, the extent, the percentage, the amount of exposure? Is it a matter of degree?

A. I would say, but in this case to agree can be a very serious disagreement.

In my case, for instance, about the evidence that I was going to discuss now, I arrived at the figures of death around the Connecticut nuclear plant that are some one thousand to ten thousand times greater than the Commission would adopt.

BY THE COURT: Go ahead.

BY MR. KACHINSKY: [Continuing]

Q. ***Okay, now, I would like to show you Exhibit No. 12, "Cancer Mortality Changes Around Nuclear Facilities in Connecticut." Could we refer you to the exhibit? Could you tell us what that exhibit is?***

A. Yes, I have it in front of me.

BY THE COURT: Just take off and tell us about it. Don't wait for him. It makes it a lot easier.

BY MR KACHINSKY: [Continuing]

Q. ***Are you the author of this paper?***

A. All right. This is a study entitled "Cancer Mortality Changes Around Nuclear Facilities in Connecticut," and this is testimony which I presented at a Congressional seminar on low-level radiation February 10th, 1978, in Washington, D.C.

BY THE COURT: And it was the truth then and you think it's the truth now?

A. I certainly do.

PERCENT CHANGE IN CANCER MORTALITY RATE (1958-1968) WITH DISTANCE AWAY FROM SHIPPINGPORT NUCLEAR REACTOR

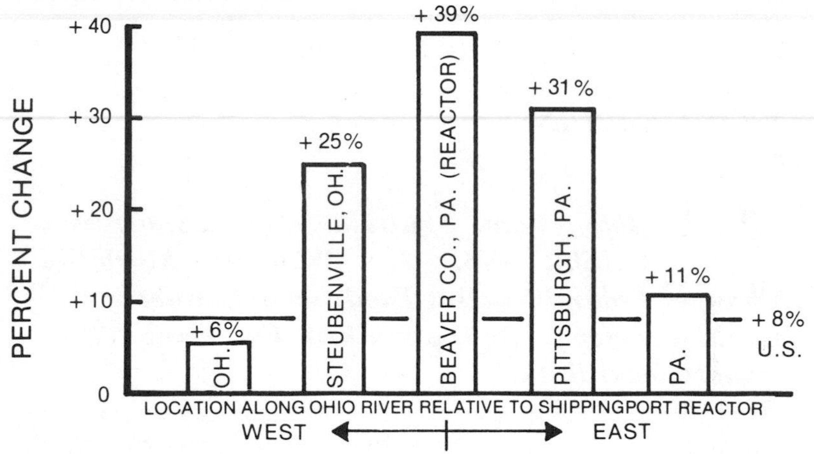

PERCENT CHANGES IN MORTALITY RATES: LEUKEMIA AND OTHER CANCERS OF LYMPHATIC AND HEMAPOIETIC SYSTEM RELATIVE TO 1959-61, BEAVER COUNTY

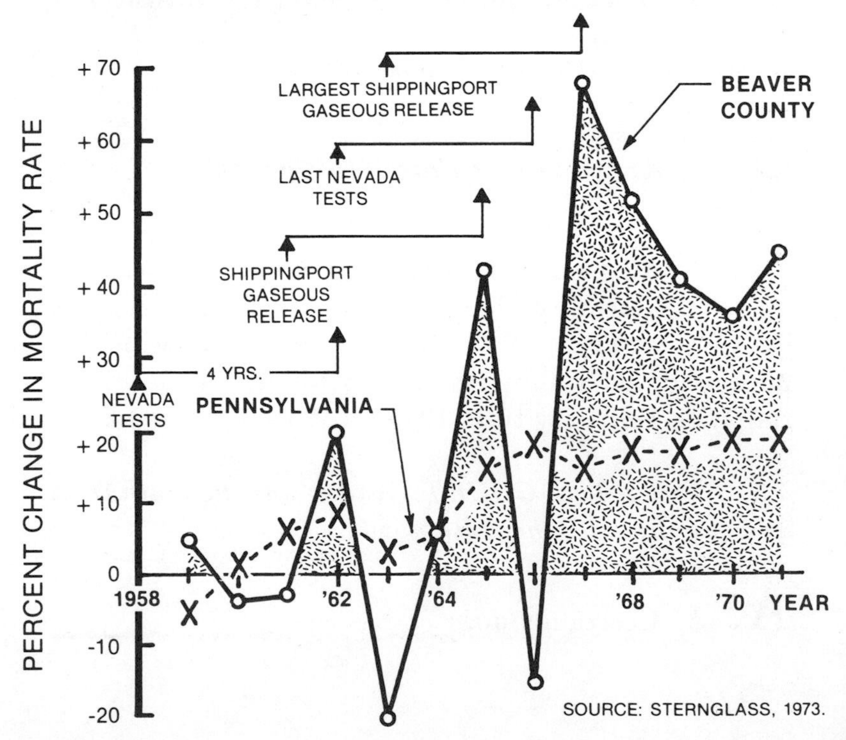

PERCENT CHANGE IN CANCER MORTALITY RATE RELATIVE TO 1959-61, FOLLOWING RELEASES FROM SHIPPINGPORT NUCLEAR REACTOR.

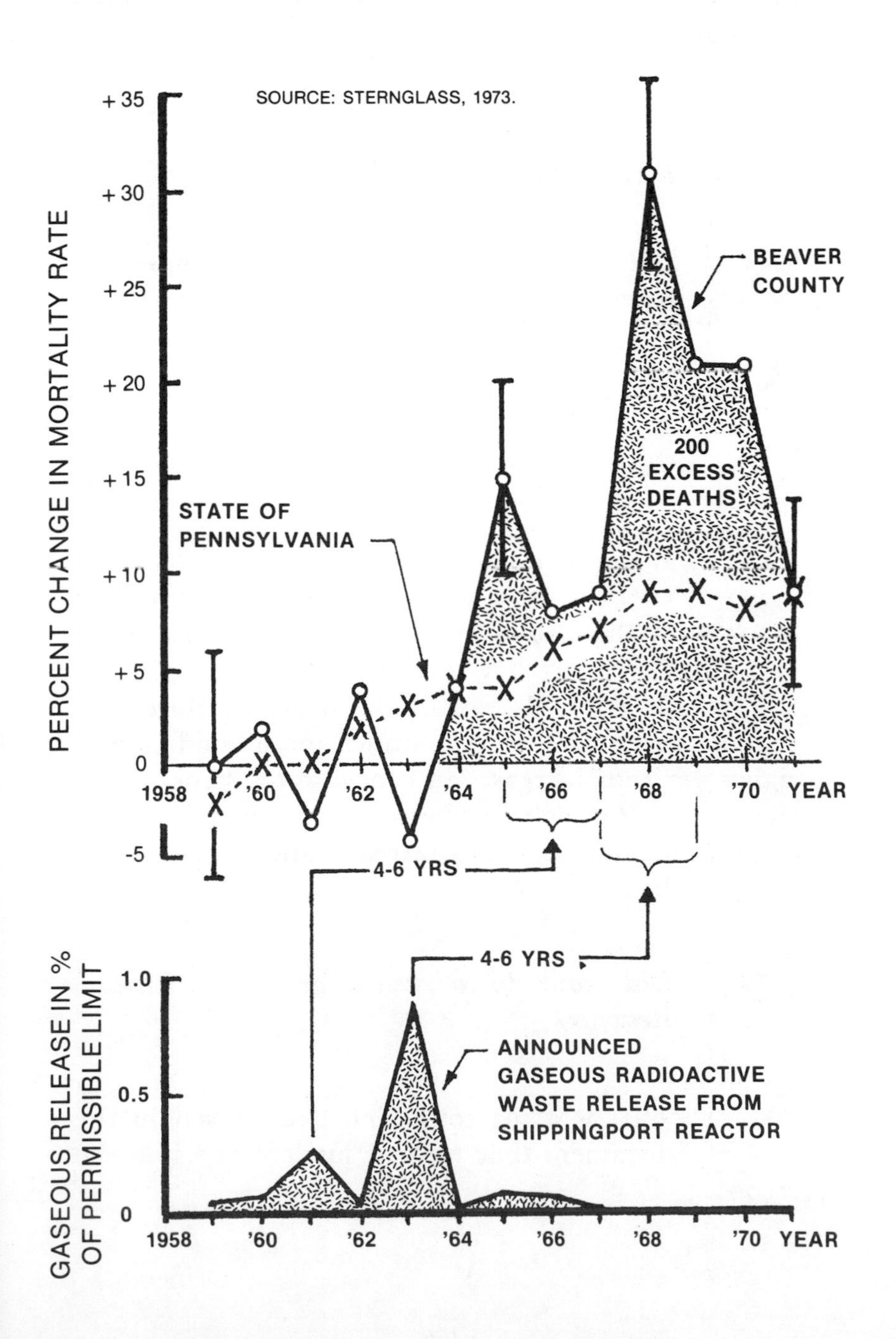

BY THE COURT: Let it be filed as an exhibit, and I will read it when we get through here.
Go ahead. Pass on to something else.

[***Marked and filed Exhibit No.12 in evidence***]

BY MR. KACHINSKY: [*Continuing*]

Q. ***Did you do a study on the Shippingport reactor?***

A. Yes, I did.

Q. ***And what did that study show?***

A. It again indicated that the number of cancer cases in the counties surrounding increased much more than would've been expected on the basis of projection carried out by the Commission.

Typically I found rises of the order of thirty per cent of cancer rates in the counties and a hundred and eighty per cent increases in cancer rates some seven, eight years after the plant began operating in the town of Midland one mile downstream and drinking the water.

Q. ***Did you do a study at the Millstone Reactor?***

A. That is a reactor which is discussed in the document that we had just entered into the record.

PERCENT CHANGE IN CANCER MORTALITY WITH DISTANCE FROM MILLSTONE NUCLEAR PLANT BETWEEN START-UP IN 1970 and 1975

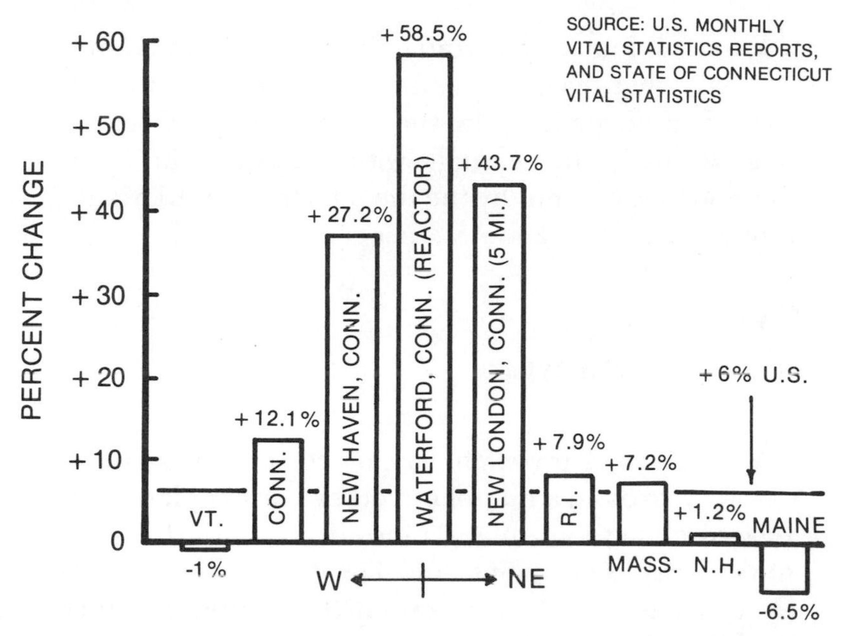

STRONTIUM-90 LEVELS IN MILK AT VARIOUS DISTANCES FROM THE MILLSTONE NUCLEAR PLANT. JULY 1976

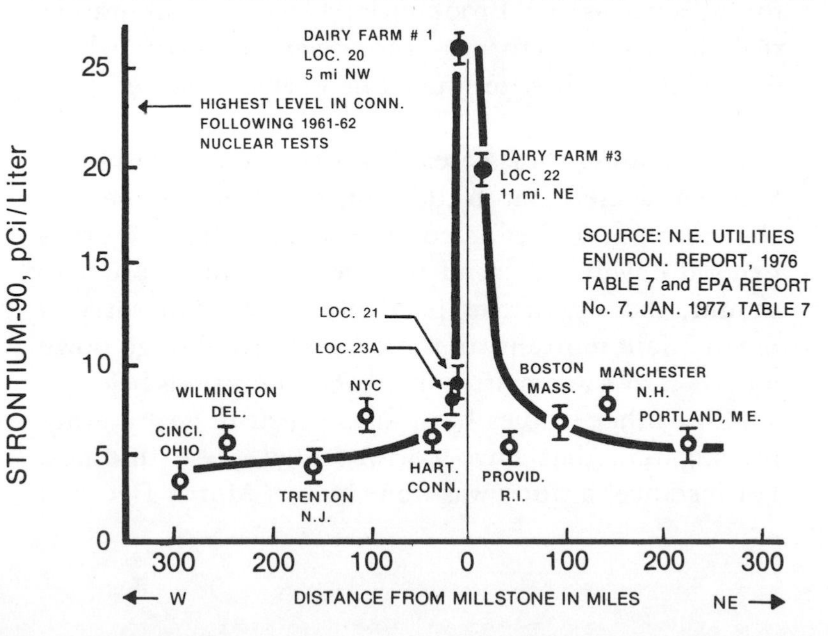

Q. ***Okay, do you know...***

A. I found essentially similar findings around that reactor. Again, a significant increase in cancer rates, greatest in the town where the reactor was located, decreasing with distance in every direction away, until by the time one gets up to Maine, it had declined by about six per cent.

Q. ***Do you know of any other studies of a similar type?***

A. Yes, I myself have carried out other studies around a number of other nuclear plants that related changes in infant mortality and birth weight to the gaseous emissions, and I've done something like about seventeen—I have examined seventeen facilities, something of that order, seventeen or eighteen, research reactors, larger nuclear plants and laboratories, such as the Brookhaven National Laboratory reactor, the reactor at the original plant where plutonium was first produced in Washington and Oak Ridge.

And in all of these cases, I was confronted with data which indicated that in the county around the reactor, the cancer rates began to reverse their trend from a general downward trend for the state and started to mount, that is, in this particular case—I'm sorry, I mean infant mortality in these cases started to go down and then rose again after the reactor emissions began.

Now, other studies have since been done by other investigators that have corroborated these findings. For instance, a study was done by Dr. Morris DeGroot of the Department of Statistics at Carnegie Mellon University, published in the proceedings of a Berkeley Symposium on Mathematical Statistics and Probability, in which he examined four nuclear reactors and their emissions and correlated them with the ups and downs in infant mortality in that area.

And he concluded that for three of the facilities, the Dresden boiling water reactor, the Indian Point pressurized water reactor and the gas-cooled nuclear reactor in Brookhaven, there was a significant, although small, correlation between the known and announced releases of radioactive gases and the actual changes in death among babies in that area.

For one reactor, the pressurized water reactor at Shippingport, he did not find a positive correlation between the releases and the change in infant mortality in Beaver County.

Subsequently, we discovered at hearings held by the NRC, which I personally attended, related to the Beaver Valley reactor, that the releases that were officially reported were claimed to have been zero releases in one year, and yet one of the operators under sworn testimony admitted under questioning by the counsel for the City of Pittsburgh, Mr. Brandon, that there were uncontrolled releases and leaks allowed from the yard, which were never reported to the authorities, and this would explain why Professor DeGroot could not find a direct positive correlation between admitted releases and the deaths of babies in Beaver County.

Q. ***Do all these studies that you just talked about, do they rely on actual deaths for their figures?***

A. That's right. I do not rely upon projections. I do not rely upon any linear hypotheses or quadratic or superlinear hypotheses. We rely solely upon a comparison between a city which has suddenly had nuclear power operating in it, before then it had a coal plant or an oil plant in it and afterwards, suddenly, for no other explainable reason, that has yet been advanced by anyone in the industry or the Atomic Energy Commission, or the NRC, there was a sudden and unexplained rise in infant deaths, in leukemia, and many years later in various types of cancers.

Then when in some cases a plant was shut down or modified or the fuel rods changed when they were leaking, there was a drop in infant mortality.

And in the case of Dresden, I have seen this twice. When the fuel rods were renewed and failed to leak, then infant mortality declined again.

When they began to leak again, the infant mortality rose again.

And when they repaired it, it declined again.

And, therefore, I believe, to the best of my scientific judgment, it is much more probable than not that these deaths are directly attributable to the operation of these nuclear plants.

Q. ***What figure does the Nuclear Regulatory Commission use for their average nuclear power plant releases? Are you familiar with the figure that they commonly use?***

A. Yes, I believe it's somewhere around a millirem or less. This is what the—that's what had been reported to the National Academy of Sciences, and it's on that basis that I believe the

National Academy in the 1972 BEIR Report thought that nuclear energy may be possible without exceeding the genetic burden on the population.

Q. ***Do you feel that that is the correct figure?***

A. No. I now have reason to believe that for many reasons that these statements were grossly optimistic and that, in fact, the actual doses to the members of the population living near these plants are hundreds of times, if not more, greater than the National Academy was told or than we had calculated in the earlier years of nuclear optimism.

Q. ***Are you familiar with the work of the Japanese doctor, Dr. Ichikawa?***

A. Yes, I am.

Q. ***Could you comment on his studies?***

A. Yes.

BY THE COURT: Does he bear you out?

A. No, sir, it is an entirely different kind of study. I will just mention it briefly. It does not relate to actual deaths. It has to do with the effect of small amounts of radiation on a flower, which is called *Tradescantia* or spiderwort, and what he was able to show is that very small amounts of radiation down to two hundred and fifty, three hundred millirems are able to change the genetic character of this plant so that it changes color, the stamen changes color, and so they have planted them around nuclear plants in Japan. They found that, indeed, the effects on these plants show genetic damage at levels well below the five hundred millirem maximum, which is presently permissible, and that this has now been found in a series of nuclear plants in Japan, and in that sense it substantiates my concern that the true doses in the environment have indeed been underestimated.

Q. ***What can you say about worker exposure rates in the United States?***

A. Well, I have examined this question to some degree, and my principal concern arises from the fact that worker exposure has also been rising faster than we had hoped in the beginning of the nuclear age, and that again this is an involuntary and unanticipated kind of exposure which, apparently now is likely to contribute significantly to the genetic burden, and, therefore, defective children in the future merely from the men who are exposed within the plant.

The figures that I have seen and estimates which appear reasonable to me in my personal experience seem to show that the projected dose just from workers alone would be equivalent to giving the whole population something like twenty to twenty-five per cent of background radiation due to the fact that they marry other people not working in the plant and have children. This would then result in an additional burden of defective children in future generations and additional death of significant magnitude, and, therefore, not only the general population, but also the worker exposure contributes to the problem of these unanticipatedly high health impacts.

Q. ***I would like to show you Exhibit No. 13 and ask you to comment on it, identify it and comment on it.***

A. Yes, this is a document entitled "Estimate of the Cancer Risk Due to Nuclear-Electric Power Generation, October, 1976, U.S. Environmental Protection Agency, Office of Radiation Programs, Technical Note ORP/CSD-76-2."

BY THE COURT: It is a government release, is it?

BY MR. KACHINSKY: Yes, it is a government release.

A. You can have a look at it.

BY THE COURT: After it's marked, pass it back to counsel, defense counsel, and let them look at it. You know what it says, you can testify while they are looking at it.

A. Yes, sir.

BY THE COURT: Go ahead and make your comments.

A. Right. This report actually deals with the estimate made by some EPA scientists independently of the likely radiation exposure as a result from the nuclear fuel cycle, including, of course, the mining and milling and the nuclear reactor releases themselves.

Their basic conclusion is that they would expect roughly about 1.1 deaths per year per operation of a single one thousand megawatt plant in the United States.

Now, I fundamentally agree with this general approach and method of calculation as far as it went at the time, but since then there are a number of things that have come to light which would simply cause me to increase somewhat this estimate.

I would say that there are two major things that cause me to believe that this is now underestimated.

On the one hand, we now have evidence that we have underestimated the amount of radiation that gets out, that gets through the food chain and gets into the human body.

Dr. Gofman already alluded to some of that in his testimony, namely that, for instance, the plutonium has only recently been found, as published in the last month or so, that a thousand times as much may go through the intestinal tract into the bloodstream as compared to what was used when this report was done.

Another similar report has come to me in the last few months by fourteen scientists at the University of Heidelberg, who were independently asked to examine the transmission of radioactivity from the nuclear power plant of a pressurized water type in Germany similar to the American design, and to evaluate the factors in the calculations that were used by the U.S. Atomic Energy Commission, the German Atomic Energy Commission, the British Atomic Energy Commission and even the International Commission on Radiation Protection.

And what they discovered is a very, very disturbing finding.

These previously uninvolved scientists concluded that the doses that were calculated for the people living around the reactor were underestimated anywhere from one hundred to a thousandfold, and that this happened as a result of an arbitrary choice of parameters in the literature which, apparently, were carried out by the International Commission on Radiation Protection and all of the other organizations in the sense that when there were about five or ten different experiments published as to how much strontium 90 gets, say, from the soil into the plant, they would always pick the lowest value, and they would pick the lowest value for how much gets from the plant into the meat of the beef cattle, and they would pick the lowest value for the factor that would determine how much would be given to a critical organ of the human body.

And by the time you multiplied all these things together, they concluded that the true doses from operating nuclear plants all over the world are probably too high by something like ten to a thousandfold.

So this is one reason why I believe that this estimate here is grossly misleading and underestimates, although in principle I subscribe to their general approach.

The second reason is that since this was done Dr. Mancuso's study has been published.

Now, up until now it was possible always to argue that a long protracted exposure at low levels from the environment, such as we get from nuclear reactor and fuel cycle releases, would be less harmful than medical exposure.

But Dr. Mancuso's data, and other recent findings independent by him, such as that of Dr. Petkau and the Canadian Atomic Energy Commission and many others now show that protracted exposure, far from being similar to a short medical X-ray. is perhaps twenty, fifty or a hundred times more serious for every millirad which is given to the individual.

Now, when you combine the greater transmission through the food chain with a greater transmission to a critical organ and then a greater risk for a given amount of dose to the critical organ, then you end up with the situation where a total health impact of nuclear power may have been underestimated, and that is their term, the German study's term, as much as ten thousand times.

BY MR. KACHINSKY: I move to have this exhibit placed in evidence.

BY THE COURT: Granted. Exhibit No.?

BY MR. EILPERIN: Your Honor, that's not a Nuclear Regulatory Commission document. It's not an admission. It's an EPA report.

BY THE COURT: Oh, I will let everything in. If he wants to file a newspaper, I will let that in, too.

BY THE CLERK: For the record, Exhibit No. 13, a report on cancer risk, dated October, 1976.

BY MR. KACHINSKY: [*Continuing*]

Q. ***I would like to show you Exhibit No. 14 and ask you to identify it.***

BY THE COURT: What is it now? Let's don't be...

BY MR. KACHINSKY: This is a translation by Dr. Sternglass on a German study, which he, born in Germany and is his first language, he translated it.

BY THE COURT: What does it show?

A. What I have just mentioned. It's part of the Heidelberg study.

BY THE COURT: Let it be filed as an exhibit. You don't have to ask him any questions about it. He has already explained it.

BY THE CLERK: For the record, Exhibit No. 14, an English translation by Dr. Sternglass.

[***Marked and filed Exhibit No. 14 in evidence***]

BY MR. KACHINSKY: [*Continuing*]

Q. ***In scientific judgment, would you say that the proceedings of the Nuclear Regulatory***

Commission have been a forum for a fair discussion and resolution of scientific issues?

A. No, sir.

Q. ***Do you think the public has been given a fair and accurate picture of the dangers to health inherent in nuclear power?***

A. No.

Q. ***In your estimation as a scientist with expertise in radiation, is the plaintiff, Jeannine Honicker, actually damaged by the nuclear fuel cycle?***

A. There's no question in my mind that there would be an increase in her risk, both to herself, both to her children and their children for generations to come.

Q. ***How many years have you studied the nuclear fuel cycle?***

A. Would you repeat that?

Q. ***How many years have you studied the nuclear fuel cycle?***

A. I first became involved in studying the effects of fallout radiation back in 1961; that would be seventeen years.

Q. ***Have you studied any of the economic aspects of the nuclear fuel cycle?***

A. Yes, I have.

Q. ***Do you think that a shutdown of nuclear power will cause widespread economic damage?***

A. No, I do not, and the reason is the following: I worked for many years at the Westinghouse Research Laboratories. I reported to a man who built the very first nuclear submarine reactor in the United States.

I participated later on in the development of the instrumentation for nuclear reactors, and I know of firsthand knowledge that the problems of nuclear energy technology and safety are so difficult that we simply have to find a way to generate our needed electricity some other way, and I did some studies as a result of this when I became convinced in the last two or three years that we would have to find a way to shut these reactors down without economic harm to the society.

I discovered in a study published by the Atomic Energy Commission itself that nuclear reactors can be converted to fossil fuel, to gas, to oil or coal at a minimal cost that, in fact, does not exceed something like fifteen to twenty per cent of the total invested so far, so that it would not be necessary to abandon our existing nuclear reactors and shut them down. All that will be necessary is to take a torch, cut off the supply line from the nuclear steam boiler and substitute a gas boiler for something less than ten to fifteen per cent of the total cost of the entire plant.

BY THE COURT: What does the normal plant cost, Doctor?

A. I have the figures with me, Your Honor, and these figures are figures which are prepared by the Atomic Energy Commission. If I may take a moment to find these figures.

BY THE COURT: Yes. I have never heard of what one costs. I'll be glad to hear it.

Now, these figures you are going to give us are the final compiled figures, not estimates, is that correct?

A. I'm sorry, sir?

BY THE COURT: Are these figures the actual cost figures or are these estimates?

A. These are actual cost figures prepared by the Atomic Energy Commission, sir.

BY THE COURT: All right. Okay.

A. Excuse me, I was not prepared for this line of questioning.

BY THE COURT: Well, it probably isn't material. It might be, but at least I will learn something.

A. I will find it in a minute. It's just—I'm sorry, bear with me. I didn't give you the study, did I?

MR. KACHINSKY: [Continuing]

Q. ***Not to my knowledge. We don't have it.***

A. That's part of it. If I can use that, that's one part, but the cost figure is what I need. Oh, yes, here it is.

First, I would like to establish the fact that such procedures have been carried out in the past, that nuclear plants have been converted by the Atomic Energy Commission when they began to leak too much, and the conversion did, indeed, save the utility the major investment, and the evidence is contained in this document entitled *Annual Report to Congress of the Atomic Energy Commission for 1968*, January, 1969.

I will read from page 97. It's under the subject heading of "Project Adjustments in Terminations."

It reads "Pathfinder. The Northern States Power Company announced plans in September, 1968, to install gas-fired boilers at the Pathfinder Atomic Power Plant in Sioux Falls, South Dakota, after a plant shutdown in September 1967, cracked and broken internal equipment in the steam system was discovered. Installation of the boilers would permit use of the turbine generators for generation of electricity, even though the nuclear steam supply system is not in operation."

Now, the other document on which I relied is entitled *Trends in the Cost of Light-Water Reactor Power Plants for Utilities,* May, 1970, Division of Reactor Development and Technology, U.S. Atomic Energy Commission. It's entitled "WASH-1150." The author is, I believe—now, I don't find the name of the author here, but it's based on information prepared by United Engineers and Constructors, Incorporated, for the Atomic Energy Commission under AEC Contract No. AT30-1-3770.

Now, the gist of this is contained on page 24, Table 2, where he talks about updated LWR power plant costs estimates, and in fact, it gives a detailed breakdown of the cost by structure improvements, reactor plant equipment . . .

BY THE COURT: Could you just give me the bottom line?

A. Right. The bottom line in this particular case, using the high escalation rate in terms of interest rates, is two hundred and four million seven hundred thousand dollars for the entire plant.

Now, in the same report, in Figure No.—in Figure 15 on page 34, there is a breakdown of the nuclear plant costs by major component and also that of a coal plant, and the boiler of a coal plant on this graph is indicated at only twenty-five million dollars.

So, when you add—you know, you build a nuclear plant, say, for two hundred and four million dollars, and then you add a fossil fuel boiler, something of the order of twenty-five million dollars, I mean at those costs, that represents a fraction of only twelve per cent of the total investment.

Now, from my study . . .

BY THE COURT: Twelve per cent?

A. Twelve per cent of the total cost of the nuclear plant.

BY THE COURT: Two hundred and four million dollars, ten per cent of that is twenty million.

A. That's right.

BY THE COURT: But all they do—you don't have to have any labor to take out the other?

A. No, that's already figured into the cost of the boiler, installation and the interest rates and everything.

BY THE COURT: Okay.

A. And so what happens here is, apparently, that if you need to convert a plant, you can save a majority of the equipment. You simply attach another boiler.

This has been done in another plant in the United States, and that is mentioned in the Petition of Jeannine Honicker in a table which also mentions this particular plant, the Sioux Falls, South Dakota plant.

It has also been done in Sweden.

Now, this additional investment—of course, everything has proportionately increased in cost since this report was written back in 19—May, 1970, but everything has gone up more or less proportionately, and I would—it's my best judgment based on my engineering experience in this field, that it could be done for certainly under twenty per cent of the total cost of the plant.

Now, you can recover some of that twenty per cent as follows, because the experience has shown that coal plants operate more reliably. All of the technology is less complex. It's possible to run a coal plant at more efficiency and generate electricity more of the time. You don't need as many shutdowns.

Typically, a coal plant, the operating record is something like in the neighborhood of seventy to seventy-five per cent or so of operating availability.

Nuclear plants, because of their greater complexity and the greater concern about leakages and the fact that it is a new technology, unfortunately has required more frequent shutdowns, that means that the nuclear plant has been operating more like fifty-five or sixty per cent of capacity.

So that in a matter of a few years, it is possible to recover economically the cost of conversion to gas, to oil, to coal or gas from coal or any of these alternatives that we now have, and we have five hundred years worth of coal, and we are making great progress in coal gasification and other things so we can put clean gas into a close metropolitan area like Connecticut or Nashville or any metropolitan area, generate the electricity without having to shut down this city and recover the cost over a period of five or ten years after the conversion.

Thank you.

BY MR. KACHINSKY: I have no other questions. Excuse me, I do.

BY MR. KACHINSKY: [Continuing]

Q. ***I just want to ask you one last question, a general question on your concern for humanity...***

BY THE COURT: Oh, no, we are—I have as much concern for humanity as anybody.

BY MR. KACHINSKY: Okay, I have no further questions.

BY THE COURT: Come around.

CROSS - EXAMINATION

BY MR. EILPERIN:

Q. ***Dr. Sternglass, you mentioned at the beginning of your testimony some studies and reports that you did in the 1960's, is that correct?***

A. That I did?

Q. ***Yes.***

A. When?

Q. ***In the 1960's on some Japanese studies?***

A. Yes, the first paper I published relating to the effect of radiation on the developing fetus and congenital defects, infant mortality and leukemia, I believe I gave my first paper on that subject, leukemia, in my Congressional testimony in 1963 and in my article that appeared in *Science* in June of 1963, that's right, yes, early '63.

 ***And you did other reports in that time period as well as in the 1960's?***

A. Later, in 19...

BY THE COURT: The answer is yes, go to the next question.

A. The answer is yes.

BY MR. EILPERIN: [*Continuing*]

Q. ***Did the BEIR Committee Report of 1972 review the reports that you had done up to that time?***

A. As a matter of fact, the answer is yes.

Q. ***All right, thank you. Do you recall this statement in the BEIR Committee Report about your reports: "The evidence"—and I will*** *quote it to you—"The evidence assembled by Sternglass has been critically reviewed by Lindop and Rotblat and Tompkins and Brown. It's clear that the correlations presented in support of the hypothesis depend on arbitrary selection of data supporting the hypothesis and ignoring of those that do not. In several regards the data used by Sternglass appear to be in error.*

"One of the most vital assumptions in the model, that without the atomic tests the infant mortality rate would have continued to fall in a geometrically linear fashion is without basis either in theory or in observation of trends in other countries and other types. The doses of strontium 90 used in the experiment referred to by Sternglass were on the order of one hundred thousand times greater than those received by humans from all of the atomic tests and were associated with extremely small differences in infant mortality, 8.7 per cent in the irradiated versus 7.5 per cent in the control mice.

"In short there is at the present time no convincing evidence that the low levels of radiation in question are associated with increased risk of mortality in infancy. Hence, for the purposes of this report, no estimate of risks are considered to be applicable." Do you recall that statement?

A. Not only am I familiar with it, but I would like to be able to explain the nature of the statement in the light of some of the things that have since happened.

Q. *Well, you can do that on redirect.*

A. May I do it now?

BY THE COURT: No, I'm not interested in your explaining it. Move along. Let's go on to something else.

BY MR EILPERIN: [*Continuing*]

Q. ***Dr. Sternglass, you also referred to a new study that you had done regarding the Millstone plant, is that correct, and that study was done at what time?***

A. The study was done last year.

Q. ***Are you aware of the fact that the Environmental Protection Agency has reviewed your study?***

A. That's right, and I disagree with the review of the Environmental Protection Agency.

Q. ***I am sure you do. May I read to you...***

BY THE COURT: Now, wait a minute, let's not—you get off the bandwagon, too. We don't editorialize here. We ask questions and get answers.

BY MR. EILPERIN: That's right.

BY THE COURT: All right.

BY MR. EILPERIN: [*Continuing*]

Q. ***Do you recall this statement by the Environmental Protection Agency: "After careful review of this report"—referring to your strontium 90 levels of milk and diet in your Connecticut nuclear power plant. "EPA's main conclusion is that the data do not support Dr. Sternglass's contention that operations of the Haddam Neck and Millstone Point nuclear power reactors have contributed to significant levels of strontium 90 and cesium 137 of milk in the vicinity of these reactors."***

A. I have recently been made aware of the study. I have examined it, and I have concluded that they have ignored the large concentration of strontium 90 in the immediate neighborhood of the stack which drops off in all directions away in direct correlation with the cancers, and they have in no way been able to explain away either the enormously high strontium levels at that plant or the changes in cancer rates, which they did not even address in the reply.

Q. ***Was your report also reviewed by Dr. Marvin Goldman?***

A. I have seen a report by Dr. Marvin Goldman in which he used essentially the same methods in order to try to discredit these findings that were used by the EPA.

BY THE COURT: And who is Dr. Goldman and why would he want to discredit your report?

A. Dr. Marvin Goldman is familiar to me as a scientist who has on repeated occasions come to testify on behalf of the utilities at various hearings where I have appeared on behalf of the citizens groups and intervenors.

BY MR. EILPERIN: [*Continuing*]

Q. ***May I ask you this, is Dr. Goldman a member of the National Council of Radiation Protection, do you know?***

A. As far as I know, he may be.

Q. ***Do you know if he's a member of the New York Academy of Sciences?***

A. As far as I know, he may well be.

Q. ***Do you know whether he's the recipient of the Atomic Energy Commission's E. O. Lawrence Award for his contributions to the understanding of the effect of bone-seeking...***

A. That is right, and the Atomic Energy Commission always awards these grants to the people who do its bidding. (Applause)

BY THE COURT: I will clear this courtroom if that happens one more time. We are not here in connection with any popularity contests. We are here in connection with a legal matter, and that's all we are here for.

If you want to have a popularity contest, go out in the street.

Go ahead.

BY MR. EILPERIN: [*Continuing*]

Q. ***Okay, Dr. Sternglass, I think you referred to the cost of a nuclear power plant as two hundred four million dollars?***

A. That was at that time the price given by the AEC.

Q. ***That was in 1970?***

A. That's right. Since then all prices have gone up, as I indicated, but the fraction of the total cost that would be used by the boiler, obviously, would go up in proportion to the total inflation rate.

Q. ***Do you know what the present cost is of a nuclear power reactor?***

A. Oh, nuclear power plants today go anywhere from six hundred to a thousand million dollars, and I suppose boilers also have gone up proportionately.

Q. ***You referred, I think, to—well, how large is the typical nuclear power plant that is constructed these days?***

A. Today in the neighborhood of eight hundred to twelve hundred megawatts.

Q. ***You referred to a Pathfinder power plant that was converted in 1968. I think you said...***

A. That's right, it was a hundred and fifty megawatt plant. The size of the plant has no material impact on the possibility of converting any nuclear plant of any size to a different fuel, just as we have recently been converting some of the plants to, say, for instance, coal.

Q. ***You referred, I think, to converting power plants to clean gas?***

A. That's right.

Q. ***Are you aware whether or not there is a sufficient gas supply to convert thirteen per cent of the electrical power to gas?***

A. Yes, I have recently examined a statement that was, in fact . . .

Q. ***The answer is yes?***

A. The answer is yes, and the fact is, to substantiate it, I would like to add that a statement was recently entered into the Congressional Record in connection with the economics of various possible fuel sources which indicated that generally the gas industry underestimates its potential and generally finds more as the economic incentive is increased, as, for instance, right now by the deregulation of gas prices.

Q. ***There was a gas shortage last year, was there not?***

A. It was largely generated by, many people believe, partly by an industry that wanted to see the gas prices go up and the deregulation of gas.

Q. ***Thank you Dr. Sternglass. No further questions.***

BY THE COURT: Step down.

Now then, let's have a little conference. Step down.

Now, I have the picture. Step down.

Now, we have here is whether or not—what we have—I have a real serious question concerning my jurisdiction. I have heard enough to know what this lawsuit is about.

I want you to talk to me, young man, and show me what kind of jurisdiction I have.

BY MR. KACHINSKY: May it please the Court, we believe that this is a Constitutional question, and our expert witnesses have shown that plaintiff, Jeannine Honicker's body has been damaged as a result of the nuclear fuel cycle.

BY THE COURT: Let's assume all of that. Now, let's go. Tell me what my jurisdiction is.

BY MR. KACHINSKY: We believe you have jurisdiction under 42 USC 1983 Civil Rights, 28 USC 1331 and 1361, and that this action is not an appeal of an administrative proceedings; that this is a direct appeal, a direct original jurisdiction civil rights case in that the plaintiff's constitutional rights have been violated. She has been afforded . . .

BY THE COURT: I take it you concede, and if you don't concede it, I will have to have further proof, but I take it you will concede that there is a divergence of opinion as to the amount of radiation which will cause cancer or other deleterious effects, is that correct?

BY MR. KACHINSKY: Yes, sir.

BY THE COURT: Do you also concede that the Commission, the Regulatory Commission has had hearings, after which it has set standards?

BY MR. KACHINSKY: Yes, sir.

BY THE COURT: All right, do you also concede and this is something that maybe you have not had an opportunity to think about in view of what came out here today, but apparently in April or sometime this year this Commission has by regulation provided that standards could be attacked by a petition under the so-called rule-making authority of the Commission? Do you concede that, that amendment?

BY MR. KACHINSKY: In the normal cases it could, but we are saying this is an exceptional circumstance.

BY THE COURT: Okay. All right, young man, let me ask you this now, we have here the situation which apparently there is a substantial divergence of opinion.

There is a procedure outlined by Congress, and Congress makes the political decisions, not this Court, whereby you can raise this question before an agency to which Congress has committed jurisdiction.

And you have a right to appeal from the decision of that agency directly to the Court of Appeals, as you know.

And then you have a right to petition cert, of course, to the Supreme Court.

Congress has indicated its desire that the District Courts stay out of this controversy, and the Supreme Court has indicated right strongly in connection with the rule-making procedure in the Yankee something-or-another case decided in April of this year, that we are to stay completely out of it, and the Courts are to leave the rule-making and the procedures up to the Commission, and the Commission has made a finding of fact.

Now, when that Commission makes that finding of fact, after having exercised its rule-making authority, and when that Commission gives the citizens an opportunity to come forward and attack that finding of fact through a rule-making procedure, where does the Constitutional issue arise? You have the due process. You have the procedure that Congress has set up for you to raise that question, and Congress has said to the District Court, "Stay out of it."

Now, that is what bothers me in this matter.

So, now then, I have given you the things that bother me, and I'm going to sit back and let you address them, all right?

BY MR. KACHINSKY: All right.

BY THE COURT: All right.

BY MR. KACHINSKY: In our position with the first petition that we filed, we asked for immediate relief and showing—and using the NRC's own figures and statistics, kind of like saying, "Come on now. This is obvious. You see it. How come you are not doing something, you know, and it shouldn't take that long."

The NRC, after the thirty days just indicated that they were starting a proceeding.

So, at that point we felt that we had another remedy, which was to take this case as an equity action under the Civil Rights Act and under the Mandamus Act for this court to act immediately under the emergency situation in this case.

BY THE COURT: In other words you admit that the Regulatory Agency is initiating an examination of the problem?

BY MR. KACHINSKY: We admit they are issuing an examination under the regular procedures, but that they have denied our administrative remedies under emergency procedure, and I would like to point out that the appeal is in the Circuit Court of Appeals. They have not made any decision yet, but we still feel even in the Circuit Court of Appeals that it is a separate action and that these are essentially two separate actions, and the uniqueness and the emergency of the situation allows us this other route to pursue, because grave Constitutional rights are being violated and the plaintiff is suffering irreparable harm.

BY THE COURT: Well, I will tell you what I'm going to do. I'm going to take this under advisement.

I will let you all submit some additional briefs if you want to submit them.

I will give you fifteen days to submit additional briefs and authorities. I will take it under advisement.

If I need any more evidence, I will set it for another hearing, and we will take some more evidence at that time if I need it.

I just—I wanted to hear some of the evidence to be sure that there was no question about what the parameters established dealt with the jurisdictional question.

Sometimes it's a lot easier to have an evidentiary hearing, at least a limited evidentiary hearing to eliminate possible ambiguities in connection with parties' contentions.

So I think I have enough here now that I know that the primary issue is whether I have jurisdiction, and I—plaintiff's attorney has conceded there is a substantial difference of opinion among scientists, so

I don't have that problem.

So I will recess this matter and give you fifteen days to file any additional authority you want, and if I determine at that time that I have jurisdiction, independent jurisdiction—that's not a good word—but if I have jurisdiction over and above the regulatory agency, which I doubt, I will set another hearing and we will go from there.

All right, recess Court.

BY THE CLERK: Everyone rise, please, Court is in recess until tomorrow morning at 8:30.

[*Thereupon Court adjourned*]

Background and Perspective

"The unleashed power of the atom has changed everything save our modes of thinking, and we thus drift toward unparalleled catastrophes." —*Albert Einstein, 1879-1955*

The testimony presented on October 2 was, as Judge Morton pointed out, a partial evidentiary hearing to determine whether he had jurisdiction to hear the case at all. In addition to the information brought out at the trial, Judge Morton also had other scientific evidence before him as a part of the record. An overview of that evidence is presented here.

Ever since the dawn of man, radiation in the natural background has been responsible for damage to health, fatal cancers and birth defects. About five hundred years ago, this damage was first observed among the pitchblende miners in Saxony and Bohemia. After five or ten years of underground mining, they began dying early of a so-called 'mountain illness.' Many years later, following Marie Curie's work with radium and Becquerel's discovery of ionizing radiations given off from natural uranium, the reason for these early deaths appeared: the radioactivity emanating from uranium, a component of pitchblende, was causing cancer. The precise manner in which these emanations lead to the development of cancers remains a topic of scientific research and debate; but, because of the experience of the uranium miners and others, the effect is well established.

After the discovery of X-rays in 1896, a number of other effects of radiation were observed. Persons who remained underneath the X-ray too long appeared with burns. Some of the earlier manufacturers of X-ray equipment sought medical aid for serious radiation burns on their hands. Lighter doses caused a reddening of the skin, either immediate or delayed. It was believed then that at low levels there was probably a threshold below which no damage occurred. As Dr. Gofman pointed out in his testimony, no one really believes that any more, but at one time it was assumed that unless you could *see* the damage as with reddened skin, any other effects were inconsequential.

As their experience and knowledge of radioactivity broadened, scientists began to consider the possibility of *latent effects*, those damages which do not appear until years, or even decades, after irradiation. Delayed effects were experienced by the Curie family and others working with radiation, and early tests with animals confirmed the link between radiation and latent cancers and leukemias, and also suggested genetic damage appearing in future generations. From these early experiences, attempts were made to establish limits on the amount of exposure to be received by individuals. For example, in 1902, the dose was set at ten Roentgens *per day*, or 8.3 rads. Such an exposure is nearly 900 times the present recommended maximum exposure of 0.17 rads *per year*. As more information emerged about the hazardous nature of radiation, the dose level considered "safe" was further reduced.

After the tragedy of Hiroshima and Nagasaki began to show the real magnitude of human latent effects, the recommended maximum was lowered still further. Until 1952, the International Commission on Radiological Protection (ICRP) suggested a "safe" value was 52 Roentgens *per year*, or about 300 times the present allowable dose.

Then studies began to reveal that even the "low" dose provided by background radiation was likely to be causing "health effects"—cancers, leukemias and genetic defects—at a certain rate per year.

In 1954, Congress passed the Atomic Energy Act, introducing atomic power for peacetime use. The experimental aspect of such an undertaking was clearly recognized by the lawmakers, who wrote in their first draft:

> "The significance of the atomic bomb for military purposes is evident. The effect of the use of atomic energy for civilian purposes upon the social, economic, and political structures of today cannot now be determined. It is a field in which unknown factors are involved."

Shortly after the formation of the Atomic Energy Commission, when atomic bomb tests were becoming virtually routine, Nobel chemist Linus Pauling and others began to investigate the effects of fallout on the health of the people. Disturbed by their findings, they began a prolonged legal battle with the Atomic Energy Commission, arguing that the fallout would cause thousands of needless cancers and premature deaths. But the cancers and deaths were not yet visible, so these early lawsuits were unsuccessful in halting the testing. World opinion eventually forced the U.S. and U.S.S.R. to sign an atmospheric test ban treaty; and, in fact, the damage predicted by Pauling and others is now beginning to become visible. On February 22, 1979, the *New England Journal of Medicine* published a study by Dr. Joseph Lyon of the Utah Cancer Registry. Dr. Lyon found that there were two and one-half times as many leukemia deaths among children born in southern Utah during the years of the atmospheric bomb testing, as compared to children born in the same area in years before and after such testing.

In April, 1979 a Senate committee investigating the cover-up in Nevada and Utah uncovered a number of documents involving former President Dwight D. Eisenhower. Informed that the tenth atmospheric test had showered fallout on unsuspecting U.S. citizens, the President had ordered the AEC to "confuse the issue" and get on with test number eleven. Dutifully, the AEC suppressed a yet unknown number of early studies on fallout, including the study of Dr. Robert C. Pendleton on fallout in milk finished in July, 1962, and the study of Edward S. Weiss finished in 1965 which linked a sharp rise in leukemia in school children to the Nevada test program. Pendleton said that funds for his studies of radon gas were cut off as a reprisal. Weiss was told by the AEC that publication of his study "will pose potential problems to the Commission: adverse public reaction, lawsuits, and jeopardizing the programs at the Nevada Test Site."

Censorship and repression was not only coming from the AEC. Dr. Karl Z. Morgan travelled to Germany in 1971 to present a radiation health paper only to find that Oak Ridge National Laboratory had wired the German authorities to destroy all 200 copies of the paper before he could give them out. Similar incidents happened to W.B. Cottrell, John Dobbs, M.W. Wexler, F.J. Davis, A.F. Gabrysh, B.R. Fish, and many other notable scientists at Oak Ridge. In 1977, Ronald Fluegge was forced to resign from the NRC when he openly disagreed with irregularities in the radiation protection and safety programs. NRC engineer Robert Pollard resigned when he learned that unsafe reactors were being licensed to operate over his protests. But as Dr. Morgan testified to Congress in 1978, "There are not many people that wish to make an issue so they will lose their job and would no longer have support for their family and no chance of getting another job in their profession. So I think what information surfaces is perhaps a very small amount of what actually goes on."

In 1973, under threat of a Freedom of Information lawsuit, the

AEC released documents showing a cover-up of a 1964 reactor safety study. The 1964 study had estimated that a serious nuclear accident could kill 45,000 persons, injure 100,000, and destroy more than 17 billion dollars worth of property. Moreover the report concluded that residual radiation could contaminate "an area the size of Pennsylvania."

In 1972, the National Academy of Sciences published the findings of its Committee on the Biological Effects of Ionizing Radiation (BEIR). According to the BEIR Report, between one thousand and nine thousand annual cancer mortalities in the United States can be expected to result from natural background radiation alone. While recognizing that an increase in the background level will increase the number of expected cancers, the BEIR Report said that, because radiation is thought to be a valuable tool to science, a cost/benefit analysis would be the recommended approach for broadening the emerging nuclear industry. Health effects would simply be noted as one of the costs we all would pay for the benefit of atomic power.

In 1975, the Environmental Protection Agency published a report on the expected environmental effects of an expanding nuclear fuel cycle. EPA stated that the potential impact on health caused by effluents from a broadened nuclear industry would be very significant, even operating at the safety levels permitted by the federal standards. Their estimate was that several thousand deaths from cancer and genetic defects would result from routine releases of radioactivity, and from losses normally experienced in handling and transporting radioactive materials.

In the meantime, Congress had recognized the conflicting nature of the promotional and regulatory aspects of the Atomic Energy Commission. In 1974, the AEC was split into two separate agencies: ERDA (Energy Research and Development Association) for promotion of nuclear energy and weaponry, and NRC (Nuclear Regulatory Commission) for regulation to protect the public health.

During this period, the standards of radiation protection were based primarily on the BEIR report and WASH-1400 (commonly referred to as the Rasmussen Report after its principal author, Norman Rasmussen, a physicist at MIT). Shortly after publication of the report, serious criticisms of its methodology and conclusions began to appear in print. The Report's loose assumptions led to the conclusion that a person's danger of being harmed by accidental releases was about the same as the danger of being hit by a meteorite from outer space.

This simile received considerable attention in the press, but it was subsequently seen to be wholly groundless. In late 1978 and early 1979, both the BEIR report and WASH-1400 were recalled by their sponsors because actual human experience had proven them wrong. Both had underestimated the risks and effects of radiation exposures.

In 1976, the Court of Appeals for the District of Columbia ordered the NRC to assess the environmental impact of the entire uranium fuel cycle, with particular attention to the possibility of recycling plutonium from breeder reactors. In the hearings which led to a generic environmental study of the mixed oxide fuel cycle (GESMO), the NRC estimated the potential impacts of the three recycle options: the present throwaway method, the uranium-only recycle, and the plutonium-uranium recycle. The range of estimates for cancer fatalities from these three options in the GESMO report, issued in August of 1976, showed that 1,100 to 1,300 cancer mortalities and 2,100 to 2,400 genetic defects would be expected to occur from 1975 to 2000, as a result of the routine emissions of the nuclear fuel cycle.

Also in 1976, Robert O. Pohl, a physicist at Cornell University, published an article in *Science* on the effect of radon gas emitted from the waste tailings left over when uranium is milled. Pohl estimated that radon gas would cause 400 eventual cancers for each reactor fuel requirement for each year of operation. The United States has 70 operating reactors at the beginning of 1979 and some 150 more under construction. The uranium in these mill tailings will continue to break down into radon for the next 4.5 *billion* years.

The question of mill tailings has been addressed by the federal government since the 1950s. These tailings are already thought to be the cause of noticeable health problems in Grand Junction, Colorado, and other mining towns. The uranium, thorium, and radium dust and radon gas from the piles are picked up by the wind and blown over the surrounding populations. The obvious solution, to put the tailings back in the mines, was inadequate, because once the ore was crushed its volume increased considerably and it simply wouldn't fit. The amount of mill tailings already present in 1979 could bury a four lane highway across the American continent to a depth of one foot. This enormous volume of radioactive material remains piled in dry, barren hills at abandoned uranium mills. Shallow burial is now proposed in order to bring the radiation emissions down to *double* the normal background level.

In 1977, during its hearing on the licensing of the Three Mile

Island nuclear power plant in Pennsylvania, the NRC addressed Pohl's publication. Chauncey Kepford, a radiation chemist from State College, Pennsylvania, testified that the mill tailings from the uranium used to fuel a single nuclear reactor could eventually cause forty million cancers over time as the uranium decayed to radon gas. In his testimony, Kepford wondered what kind of cost/benefit ratio could exist between the temporary thirty-year burst of electricity from one reactor and the forty million potential deaths it would cause.

Dr. Walter Jordan, then Associate Director of Oak Ridge National Laboratories, had been chosen to sit on the Atomic Safety Licensing Board panel, and was Director of the board when Kepford testified. Jordan went over Kepford's calculations and found essentially that both Pohl and Kepford had been correct: potential health effects of this radon gas emission could be estimated per reactor by multiplying the amount of tailings by the length of time it will take them to break down. Jordan wrote a memorandum to James Yore, his superior at the Atomic Safety Licensing Board, stating that the NRC estimates of radon gas releases from mining and milling were in error, and that the correct value would be some *one hundred thousand times greater*. Jordan estimated that the expected deaths would be in the hundreds for each reactor for each year. In correspondence with the late congressman Clifford Allen, he subsequently refined this number to four hundred expected fatalities per reactor per year, citing Pohl.

At the same time, a study was under way at the University of Pittsburgh, under the direction of Dr. Thomas Mancuso. This study, originally contracted by the AEC, was investigating the health of the workers at the Hanford Atomic Works in Seattle, which had been in operation since the 1940's. In 1976, Dr. Samuel Milham of the Department of Health of the State of Washington found a higher rate of cancer among the workers at the Hanford Atomic Works than in the state of Washington as a whole. When Milham's work was published, considerable pressure was placed on Mancuso from the Department of Energy to publish his preliminary findings, which indicated that Milham was incorrect and that the Hanford Works was as safe a place to work as anywhere.

Mancuso objected to such an early publication because of the tentative nature of his results, and the well-known latent effects of radiation. Since cancer typically appears twenty-five years after it is actually induced by radiation, Mancuso wanted a few more years to build his data base before drawing any conclusions.

Because of his reluctance, Dr. Mancuso was then removed from his position as chief epidemiologist of the project and the study was turned over to Oak Ridge National Laboratories. Mancuso's project supervisor at The Department of Energy then left DOE to work for Oak Ridge. He became the new project director for the Hanford study. Oak Ridge's reanalysis of the data was inconclusive, and control of the study was subsequently transferred to the Hanford Works itself.

In November of 1977, Mancuso enlisted the aid of epidemiologists Alice Stewart and George Kneale of Birmingham University, U.K., and published the results of his preliminary analysis of the Hanford data. As Dr. Gofman's testimony in Federal court indicated, Mancuso's study showed a five per cent higher incidence of cancer among workers at this early stage, when exposure rates were actually lower than in more recent years. Mancuso's longer term study indicated that the risk of cancer from radiation was seriously underestimated by government standards.

In 1978, former Congressman Paul Rogers of Florida held hearings in Washington on the subject of low-level ionizing radiation. Dr. Mancuso testified to his findings at Hanford. Also there was Dr. Irwin D.J. Bross, mentioned by Dr. Gofman in his testimony. Bross had found an epidemic of cancer among patients of the National Cancer Institute's mammography program, and a high rate of leukemia and other cancers among children born of women who had received one or more X-rays in pregnancy. Dr. Bross also lost his federal funding as a result of his studies.

The Rogers hearings provided considerable evidence for the theory that ionizing radiation from the atomic program was causing cancers. Among the witnesses were a number of survivors from the Nevada atomic weapons tests. Between 1945 and 1963, these tests included marching several hundred thousand soldiers through radiation fields as a human experiment to learn how men perform under the stress of radiation bombardment. Among the men who had been at the "Smoky" test in Nevada, the rate of leukemia was over four hundred times the national average. Leukemia is the first of the latent cancers to appear.

Also at the Rogers hearings was Dr. Thomas Najarian, of the Boston Veterans' Hospital. He had attempted a study of the atomic shipyard workers at the naval submarine base in Portsmouth, New Hampshire, and other places. Although the Department of the Navy refused to co-operate, Dr. Najarian's study of death certificates indicated a very high rate of cancer among the workers on the "Nautilus" and other atomic submarines. Also at the hearing was one of the Portsmouth shipyard workers, Ronald Belhumeur, who testified that he was the only person left alive in his maintenance crew. The rest were dead of cancer or leukemia. Two of his supervisers had died within six months of each other from the same type of leukemia in 1977.

To refute the view of these witnesses at the 1978 hearing was Admiral Hyman Rickover, who at the time of the Rogers hearing testified that there was no health problem at the shipyard. Then, in January of 1979, it was revealed that Admiral Rickover had deceived the Committee by concealing a study done by the Navy which indicated a substantial shipyard problem related to radiation exposure.

Also testifying at the Rogers hearing was Dr. Edward Radford, chairman of the committee that had prepared the BEIR Report for the National Academy of Sciences in 1972. Radford testified that, based on new scientific findings since 1972, the effect of radiation exposure was probably twice as great as had been thought, and that as little as twenty-five millirems annual whole body radiation exposure would cause about a one per cent increase in the rate of cancer. The natural background rate of radiation is 125 millirems per year. A one per cent increase in cancers would result in about 3,500 additional U.S. cancer deaths per year.

Later in 1978, Dr. Pohl testified at the Black Fox reactor license hearings in Oklahoma and elsewhere that an increased rate of cancer could be expected near uranium mills and mines. Also testifying was Dr. Stanley Ferguson, who stated that, as a scientist and epidemiologist for the Department of Public Health in Colorado, he could see that there were higher rates of cancer and leukemia in the areas around the mines and mills. The rate of leukemia was two to three times what was considered normal.

In August, Dr. William Lochstet, a physicist with Pennsylvania State University, estimated that deaths from radon emitted in the government's proposed control plan for burying uranium tailings would result in fourteen million deaths from cancer and genetic defects over the hazard life of tailings from one mine.

In the same year, Dr. Ernest Sternglass published his study of the Millstone Reactor in Connecticut. He found a higher cancer rate, peaking at the plant, and *descending in all directions* out to seventy miles from the plant.

As Dr. Sternglass testified on October 2, the most recent study on radiation protection was done at Germany's prestigious Heidelberg

University, where an interdisciplinary group performed an overall evaluation and critique of current standards and standard-setting organizations. The German study showed that estimates of health effects may be too low by a factor of 10,000.

In June, 1978, Oak Ridge National Laboratory published a paper indicating that technetium-99 is one of the most dangerous radionuclides in the nuclear fuel cycle. Because nothing had been known about technetium, for many years it had been regarded by government regulations as unimportant. The Oak Ridge paper indicated technetium was 100 to 1,000 times more concentrated in food grown near nuclear facilities than had been thought, and that it was more prevalent than strontium-90 in nuclear wastes. While almost nothing is known about the biological effect of technetium, scientists now believe it is a very serious long-term soil contaminant.

In September, 1978, in an article prepared for the Department of Energy and published by the NRC as "Scenarios of C-14 Releases from the World Nuclear Power Industry from 1975 to 2020 and the Estimated Radiological Impact," *Nuclear Safety*, Vol. 19-5, p. 602, the following statement appeared:

> "The estimates of cumulative potential health effects based on integration over infinite time (effectively 46,000 years or about 8 half-lives of C-14) are as follows: 110,000 cancers and 75,000 genetic effects from the pessimistic scenario; 21,000 cancers and 14,000 genetic effects from the optimistic scenario; 22,000 cancers and 15,000 genetic effects from the intermediate scenario; 100,000 cancers and genetic effects from the C-14 formed in nature between 1975 and 2020; and 380,000 cancers and 250,000 genetic effects from the C-14 formed by the detonation of nuclear explosives from 1945 to 1975."

On September 15, 1978, Drs. R.T. Larson and R.D. Oldham published an article in *Science* that indicated plutonium is thousands of times more dangerous than estimated when carried in chlorinated water. Using Larson and Oldham's figures, and the average plutonium levels found by the State Health Department in the drinking water of Bloomfield, Colorado, health physicist Karl Morgan estimated a 30% increase of cancer risk to persons drinking from the Bloomfield reservoir or municipal water system.

On April 9, 1979, a study was released by the University of California Nuclear Weapons Lab Conversion Project which correlated a direct relationship between the increase of cancers of the tongue, stomach, ovary, brain, pancreas, and thyroid in the Bloomfield vicinity with increasing plutonium contamination coming from Rocky Flats nuclear weapons plant. The study, based on data from the Colorado Regional Cancer Center and the National Cancer Institute and conducted by Dr. Carl J. Johnson, an epidemiologist at the University of Colorado Medical School, showed that men living 13 miles downwind or generally east of the plant had a testicular cancer rate 140 per cent higher than expected. Throat and liver cancer were 60 per cent higher and rates of leukemia and lung and colon cancer were 40 per cent higher.

Overall cancer rates were 24% higher in men and 10% higher in women 13 miles from the plant. In the area 18 to 24 miles downwind, overall cancer rates were 8 per cent higher in men and 4 per cent higher in women. While the Department of Energy, which operates the plant, admits only one cancer was caused by the excess levels of plutonium in the environment, the Johnson study indicates there have been 501 unexpected cancer cases among the general population. The study is limited to only a three year period. Plutonium-239 has a hazard life of 500,000 years. Once released to the environment there is no way to reclaim the tiny particles.

On January 12, 1979, Judge Morton dismissed Jeannine Honicker's complaint against the NRC for lack of jurisdiction on the peril radiation poses to life and health. Judge Morton's memorandum concluded:

> Plaintiff insists that defendants have admitted that an imminent peril to plaintiff's health and life exists as a result of the operation of the nuclear fuel cycle, and that therefore the only question facing the court is whether or not this hazard violates plaintiff's constitutional and statutory rights. *If this assessment of the situation were accurate, the court would not feel constrained by the doctrine of primary jurisdiction and would not hesitate to act to protect plaintiff's rights.* The statements upon which plaintiff relies cannot, however, be characterized as admissions that ordinary fuel cycle activities will cause death or disease to plaintiff or any other persons. An evaluation of the facts by the NRC is thus required to determine what health risk is present. The court therefore defers to the NRC, and because any final resolution of the matter by that agency is exclusively reviewable in the court

of appeals, the court sees no reason not to dismiss this case in its entirety. (emphasis supplied)

On January 29, 1979, a notice of appeal from Judge Morton's decision was filed with the Sixth Circuit Court of Appeals. Simultaneously application was made to the U.S. Supreme Court to expedite the NRC procedure because of the life and death situation. On April 16, 1979, the Supreme Court denied certiorari. On the same day, the Sixth Circuit advanced its hearing to the earliest available date because of the immediate threat to human life. At this writing, the matter is still under consideration by the NRC and the Courts.

What Will Happen Without Nuclear Power?

What would happen if nuclear power were shut down? Suppose all the operating reactors in the United States were ordered closed just as the worst snowstorm in the century swept over the country. Suppose temperatures dropped to the lowest on record. Suppose the northeastern states, which are the most heavily reliant on nuclear power (over 25% nuclear) were to experience a sudden surge in demand for electricity. Suppose this sudden demand were not just 1,000 megawatts greater than the previous record (as has happened before) but 10,000 megawatts greater. Without nuclear power would we all freeze in the dark?

No.

Even with this worst possible situation, *there would be no region of the United States which would experience shortages.* It would not be necessary to use the national system connecting the country's regional grids to transfer power from Canada to the northeast or from the Dakotas to the southeast, as happened during the 111-day coal strike in 1978. It would not be necessary to curtail commercial lighting 20%, as was required—and accomplished without economic impact—in Los Angeles in 1973. The independent regions of the national utility grid have enough reserve in fossil-fuel generators which are now idle to provide adequate electricity in the event of a nuclear shutdown. Given sufficient notice, the utilities can take care of themselves.

How did this situation come to be?

In the present situation, seventy nuclear power plants produce about 13 per cent of America's electricity. Nuclear power is largely the creation of our economic and political system. From the point of view of a utility, which is, of course, a profit-making corporation, the more money it has flowing through its system, the better. Most states permit utilities to charge their building expenses each year to the electric subscribers in order to be sure of meeting future demand. So a utility must justify its rate increases by projecting strongly increased demand, with the resulting need for new facilities. It is now coming to light that many requests for expansion, which resulted in the construction of nuclear power plants, were based on incorrect estimates of what the actual demand load would be.

In recent years, therefore, the capacity of the electrical power industry has been growing at a much faster rate than has the actual demand. The latest figures available in *Electrical World*, the industry's magazine, show that the peak margin capacity for electrical production has been compounding steadily for the last ten years and is now 38 per cent. That means that at the peak of summer or winter usage, one power station in three sits idle because it isn't needed.

The table presented in *Electrical World* reads:

Power in million KW	1966	1967	1968	1969	1970	1971	1972	1973	1974	1975	1976	1977	1978
Capability	221.5	258.8	279.8	301.2	327.8	354.6	383.0	416.8	444.4	479.3	498.8	517.1	552.3
Peak Usage	203.9	214.0	238.6	258.3	275.4	293.1	320.2	344.9	349.3	356.8	370.9	394.9	399.9
Peak Margin (%)	18.4	20.8	17.2	16.6	18.7	20.9	19.6	20.8	27.2	34.3	34.5	31.0	38.0

United States Electrical Industry: Capability, Usage, and Margin of Excess Power, 1966-1977
Source: *Electrical World*, Sept. 15, 1977 and Sept. 15, 1978; the figures for 1978 are estimates.

Presented graphically, the same numbers look like this:

Actual and Forecasts of Peak Capability and Peak Usage, U.S. Electrical Industry, 1966 to 1990.

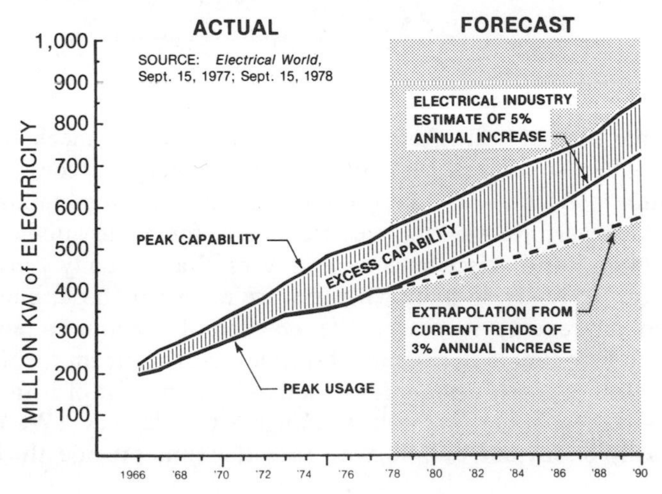

The continuation of past trends, assuming a more or less constant growth rate, shows that unneeded electrical capacity will probably increase, not decline. Removing nuclear stations from use would provide the optimum margin of reliability for electrical supply—between 15 and 25 per cent. This may indicate that utilities have been hedging the nuclear bet all along. All these figures assume little change in the nation's energy habits. But this assumption is incorrect. The country is already changing its habits. Moreover, even small improvements in the efficiency of large systems can result in drastic reductions in electric demand.

Actual and Forecasts of Gross Peak Margins (Percentage Ratio of Excess Capability to Peak Usage), U.S. Electrical Industry, 1966 to 1990.

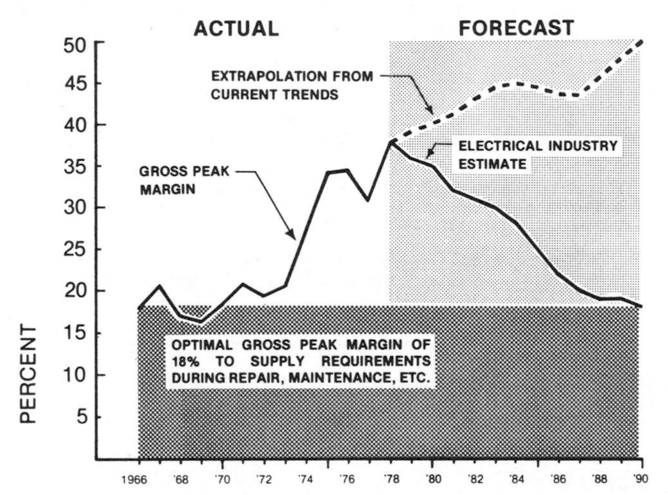

About 58 per cent of all energy employed in the United States is used in the form of heat. Another 38 per cent provides mechanical motion. The rest, about 4 per cent of delivered energy, accounts for all lighting, electronics, telecommunications, electrometallurgy, electrochemistry, arc-welding, electric motors for home appliances and railways, and miscellaneous end uses that actually *require* electricity. To really do more with less, we must match our energy needs to our energy sources. We now build mammoth power stations to heat fuel to thousands of degrees to drive steam turbines sending electric current hundreds of miles—losing most of it in the process—to heat water or small buildings a few degrees. We can more easily orient our buildings to face the sun, provide thicker walls to store solar heat through the night and, with other small changes of this type, begin to live on our energy income, instead of spending our limited resources as if there is no tomorrow.

Present energy policy is based on a choice of evils. The dominant view in policy circles is that the oil and gas resources of the world are too small to sustain our economy until transition to solar power can be accomplished. The belief is common that either we must tame the power of the atom or face the likelihood of total chaos when oil begins to run out—economic depression and nuclear war over the remaining Arab oil reserves is the picture painted by the Departments of Energy and Defense.

But nuclear electricity is unlikely to make significant inroads into markets for directly-consumed fuels—markets that now represent 90% of all energy and end-uses. Even under the most ambitious nuclear development programs, nuclear power is unlikely to supply even 10% of end-use energy 50 years from now, and *will have no effect on averting the oil crisis* according to a recent study for the U.S. Arms Control and Disarmament Agency. Nuclear power will make us dependent on costly imported uranium as well as aggravating our dependence on foreign oil. Only improvements in the productivity of energy—doing more with less—will extend the lifetime of remaining fuels beyond 2025, providing sufficient time for a transition to safe, solar energy resources without catastrophic global changes. While continuing down the nuclear path will continue to waste capital, cost jobs, and destroy human rights, a transition to safe energy will provide "growth with a human face" and at least three million additional jobs in the United States alone.

A sane energy policy must consider the fundamental assumptions its decisions are based upon, because these assumptions will also dictate the kind of society we are to have. We can live in a wasteful, armed, and frightened society, or we can live peacefully within our means. We cannot continue to weigh human lives in the same balance with electrical power. We cannot continue to squander humanity's entire heritage of natural resources in a few generations. We must choose to move in a positive direction, rapidly and decisively, if we are to fulfill our human role as caretakers of our planet and curators of life.

Conclusion

The hearing in Federal court on October 2 was the beginning of a process designed to close the nuclear industry in the United States. The defenders of nuclear power, no longer able to seriously deny the likelihood of radiation-caused cancers and birth defects, have fallen back on the thinnest of arguments.

For example, in their response to the Honicker petition, the NRC staff suggested that a cure might be found for cancer, or that genetic engineering will eliminate birth defects. So far, the few effective "cures" for cancer have proven nearly as agonizing as the disease itself, with financial ruin the price for the families of the survivors. Hospital costs and doctors' bills wipe out life savings.

Today a child born in America has one chance in three of dying of cancer in his or her lifetime. Cancer is responsible for the deaths of more children than any other known disease. Yet the American Medical Association, following economic rather than medical considerations, stoutly defends nuclear power.

Nuclear power is already random, compulsory genetic engineering. But there must be a human right *not* to be genetically engineered. The U.S. Constitution established the federal government with a power delegated from the people by their willing consent. Who has consented to give cancer to our children? Who has consented to a nation of human guinea pigs? The government abuses its authority when it licenses random, premeditated murder.

The nuclear defenders argue that fossil fuels produce more ill health than nuclear power. But this comparison is limited to the near-term, and ignores newly-developed methods of burning fossil fuels cleanly. Beyond the arbitrary 50 year cut-off point chosen by government agencies, nuclear power causes many millions more cancers and genetic effects than coal or oil. Yet the entire issue of fossil fuels versus nuclear power is an unnecessary controversy. It vanishes when we remember the enormous energy now available from non-polluting, renewable natural sources like the sun, the wind, and the tides.

Nuclear power was established as a government monopoly. The development of nuclear power runs a course contrary to the development of human rights. Recognizing this, people all over the world have come together to try to solve this common problem.

Organized labor, the scientific community, attorneys, consumers and environmentalists have joined to change the pattern of nearsighted folly. The case of *Honicker vs. Hendrie* shows one course of action.

Yet far more serious than government mistakes are the unchecked abuses of multinational corporations, answerable only to the profit motive, now expanding nuclear reactor sales around the world. Nuclear reactors, while producing electricity, also produce the material for nuclear weapons. What are non-proliferation treaties worth if nuclear weapons factories are within the grasp of every country and subnational militaristic organization? Some corporations are doing the world a great disservice.

The overwhelming majority of the world's people still find their roots in the natural world. If the peoples of the world are to develop free and egalitarian societies, they must keep nature free to support them. Modern peoples are sometimes proud of their scientific achievements; but even the most sophisticated technology rests on mere *discoveries* of the ways of nature. Ecological systems are a delicate balance of forces arrived at by a billion-year process of trial and error. With our cultural perspective of only a few thousand years, we upset that balance at our peril.

Shutting down the nuclear power industry is one step in creating a world free of fear and intimidation, where men and women can walk the earth safely for generations, and children can be born into the world, grow up and develop normally. We must selflessly abandon the nuclear option for the sake of our children. We no longer have any viable alternative.

APPENDICES

GLOSSARY

alpha rays — comparatively large, slow radioactive particles emitted from the nucleus of an atom. They are easily deflected but can cause great damage if inhaled or ingested.

background radiation — radiation (at a rate of 100-150 mrem per year) coming from space or from the earth that is not the result of man's activities.

beta rays — radioactive particles emitted from an atom that are smaller and faster than alpha rays and can penetrate several layers of tissue.

biosphere — the entire expression of life-force on our planet including the earth and all her inhabitants, the atmosphere, land-masses, and oceans.

body burden — the amount of radioactive material which lingers in the body organs.

containment vessel — the large reinforced concrete shell placed around a reactor to contain any radioactivity that might escape from the reactor itself. Soviet reactors, until recently, were built without these.

control rod — a neutron-absorbing rod which can be inserted between the fuel rods in a reactor core to slow the reaction process; when raised from the core a chain reaction is started.

core — the center of a reactor where the fission reaction occurs, producing heat.

cosmic rays — the stream of ionizing radiation from space consisting of muons, electrons, neutrons and protons which is more intense at higher altitudes and polar latitutes; a natural source of radiation.

criticality — the point at which a nuclear chain-reaction becomes self-sustaining.

critical mass — the smallest amount of a fissionable material that can sustain a nuclear chain reaction.

crud — highly radioactive corrosive material that builds up in the pipes of a nuclear power plant making routine servicing impossible; it costs $30 million to clean out the crud every ten years.

curie — a unit of radioactivity giving off 37 billion (3.7×10^{10}) disintegrations per second, the radioactivity of one gram of radium. Named after Marie and Pierre Curie, the discoverers of radium.

decontamination — scrubbing of radioactivity from people, surfaces, and equipment.

Department of Energy (DOE) — the agency created by Congress in 1977 which includes the Federal Energy Administration (FEA) and the Energy Research and Development Administration (ERDA).

doubling dose — that dose of radiation which will double the rate of mutation or death or the occurrence of a particular disease in a given population.

effluents — radioactive outflow in a liquid, gaseous, or solid form from some part of the nuclear fuel cycle or storage system.

emergency core cooling system (ECCS) — a safety system designed to immediately flood a reactor core with water following a loss-of-coolant accident (LOCA).

Energy Research and Development Administration (ERDA) — one of the two agencies (the other is NRC) created when the Atomic Energy Commission (AEC) was divided in 1975; 84% of its staff came from the AEC; controls the energy research budget and develops nuclear weapons.

enrichment — a uranium filtering process by which the naturally occurring uranium-235 is concentrated for reactor fuel use or weapons use.

environmental impact — the effect on an environment of entering a new substance or activity into that environment. For example, milling uranium leaves large amounts of radioactive tailings that give off radon gas which results in a statistically expected number of lung cancers and other health effects.

fallout — airborne radioactive fission debris created by atmospheric nuclear weapons and nuclear power plants.

fuel cycle — the sequence of steps needed for the production and combustion of fuel to produce nuclear energy including mining, milling, conversion, enrichment, transportation, and waste storage.

fuel rod — a single tube of zircaloy or stainless steel that is filled with uranium fuel pellets; because the tube is designed to a thickness of three thousandths of an inch in order to facilitate passage of neutrons, it cannot prevent leaks of radionuclides into the cooling water.

gamma rays — radioactive particles or waves emitted from an atom that are smaller and faster than alpha and beta rays and can penetrate steel and concrete.

genetic effects — those effects of radiation that are not seen in the body of the irradiated persons during their lifetime, but that are present and are transmitted to their offspring in some later generation.

half-life — the number of years required for the decay of half the radioactivity in a radioactive substance.

hazard life — the number of years over which a radioactive substance continues to emit radiation.

health effects — the effects of radiation exposure on the body including death, disease, premature aging, and genetic defects.

intervenor — the legal designation of a person or group other than the licensor or applicant for a license, who can participate in the licensing hearing for a particular power plant.

ionization — the process of adding or removing sub-atomic particles so as to form ions, which are electrically charged.

ionizing radiation — radiation carrying an electrical charge which produces ion pairs and leaves a damaging track or trail in its passage through matter.

irradiation — the act of being exposed to radiation. A person is irradiated when he or she absorbs a dose of radiation.

latent effects — damaging effects of radiation exposure which do not appear until some years after irradiation.

light-water reactor — a type of reactor that uses ordinary water instead of deuterium, sodium, or various gaseous coolants; includes boiling water reactors (BWR) and pressurized water reactors (PWR).

loss-of-coolant-accident (LOCA) — a reactor core accident in which coolant water is lost from the primary cooling system which can result in core overheating and meltdown.

Manhattan Project — a code name for the project begun in 1942 that developed the first atomic bomb that was produced at Oak Ridge, Tennessee and dropped on Hiroshima, Japan.

material enaccounted for (MUF) — fissile material that is missing at any point along the nuclear fuel cycle; since 1978, called "inventory difference."

meltdown — a serious nuclear accident in which the cooling systems in a nuclear reactor do not prevent the nuclear fuel core from melting with the potential associated release of large amounts of radiation.

National Environmental Policy Act (NEPA) — a 1969 federal law that created the Environmental Protection Agency (EPA) and that requires, among other things, that each applicant for a federal license to build a nuclear facility prepare an environmental impact statement.

neutron — one of the elementary particles of an atom which may be emitted from the nucleus of an atom and can penetrate human tissue.

nuclear fission — the splitting of atoms accompanied by the release of part of the mass into energy. It is the principle of the atomic bomb.

nuclear fusion — the combining of lightweight atomic nuclei into a nucleus of heavier mass with a resultant loss in the combined mass, which is converted into energy. The principle of the hydrogen bomb.

Nuclear Regulatory Commission (NRC) — Nuclear Regulatory Commission, an agency established by Congress in 1975 to provide regulatory controls for the nuclear industry for protection of the public health. Half of its annual budget goes to ERDA.

off-gas — radioactive gases routinely released into the atmosphere from a reactor, usually through a filtered stack.

pressure vessel — a large 400-500 ton container of welded stainless steel, usually about 8 inches thick, that houses the core of pressurized water reactors.

Price-Anderson Act — a Congressional Act that limits the liability of reactor owners and the federal government to $560 million in the event of an accident. WASH-740 estimates a nuclear accident could cause $17 billion in property damage.

rad and **millirad** — a radiation measure that refers to the **r**adiation **a**bsorbed **d**ose, which is equal to about 83% of the Roentgen value. A millirad or mrad is a thousandth of a rad.

radiation — the emission and propagation of energy through space or through matter in the form of particles or waves. Radiation emanates from atoms and molecules undergoing internal change.

radiation sickness — an illness induced by sudden, intense exposure to ionizing radiation; can range from nausea to death.

radioisotope — a naturally occurring or artificially created radioactive form of a chemical element.

radium — a decay product of uranium.

radon — an alpha-emitting radioactive gas given off by radium.

radon daughters — the radioactive decay products of radon-222: polonium-214, polonium-218, lead-214, and bismuth-214.

rem and **millirem** — a radiation measure that refers to the **r**adiation dose **e**quivalent in **m**an. The relation between rad and rem depends on the kind of particle emitting the radiation: for gamma rays, 1 rad = 1 rem; for beta, 1 rad = 10 rem; for alpha, 1 rad = 30 rem.

Roentgen — the term used for measuring ionizing radiation from a radioactive source. It is equal to the quantity of radiation that will produce one electrostatic unit of electricity in one cubic centimeter of dry air at 0° C.

SCRAM — the sudden shutdown of the fission reaction in a reactor by remote control insertion of the control rods.

shutdown — the halt of the nuclear fuel cycle as a means of generating electrical power. After shutdown, further activity of the nuclear industry would consist of waste handling and containment, decontamination, and possibly conversion of existing nuclear facilities to other methods of power generation.

tailings — radioactive sand that is left over after the milling of uranium ore; 85% of the original radioactivity is left in the tailings after milling.

thermal pollution — excess heat from a nuclear power plant that is discharged into the air and/or water of the surrounding environment.

transient — the atomic industry's word for reactor core accidents.

WASH-740 — an AEC reactor safety study that said large numbers of people might die from a nuclear accident, secretly classified for nearly a decade.

WASH-1400 — the report that white-washed WASH-740, two years after the existence of WASH-740 was revealed.

BIBLIOGRAPHY

AEC, Reactor Safety Study, WASH-740, 1964.

AEC (United States Atomic Energy Commission), Environmental Survey of the Uranium Fuel Cycle, WASH-1248, 1974.

AEC, Reactor Safety Study, WASH-1400 (Rasmussen Report), 1974.

Archer, V.E., Geomagnetism, Cancer, Weather, and Cosmic Radiation, *Health Physics,* 34:237, March, 1978.

Archer, V.E., Wagoner, J.K., and Lundin, F.E., Lung Cancer Among Uranium Miners in the United States, *Health Physics,* 25:351, 1973.

BEIR Report, The Effects on Populations of Exposure to Low Levels of Ionizing Radiation, National Academy of Sciences, National Research Council Committee on the Biological Effects of Ionizing Radiation, Washington, D.C., November, 1972.

Berger, John, *Nuclear Power: The Unviable Option,* Ramparts Press, Palo Alto, CA, 1976.

Bertell, R., The Nuclear Worker and Ionizing Radiation, Address, American Industrial Hygiene Association Meeting, May 9, 1977.

Bertell, R., X-Ray Exposure and Premature Aging, *J. Surg. Oncology,* 9:4, 1977.

Bertell, R. Testimony Relative to Human Health and Nuclear Generation of Electricity, Ros. Pk. Mem. Inst., NY 14263, March, 1978.

Bross, I.D.J., Low Level Ionizing Radiation Is Hazardous to Health, A Cover-up and Its Consequences, Testimony to the House Subcommittee on Health and Environment, February 8, 1978.

Bross, I.D.J., Hazards to Persons Exposed to Ionizing Radiation (and to their children) from Dosages Currently Permitted by the Nuclear Regulatory Commission, presented to NRC, April 6, 1978.

Bross, I.D.J., and Natarajan, N., Leukemia from Low-Level Radiation—Identification of Susceptible Children, *New England J. of Medicine,* 287:107, 1972.

Bross, I.D.J., and Natarjan, N., Genetic Damage from Diagnostic Radiation, *J. American Medical Assoc.,* 237-22:2399, May 30, 1977.

Bruland, W., et al., Radiookologishes Gutacten zum Kernkraftwerk WYHL, Universitat Heidelberg, May, 1978. (Radioecological Assessment of the WYHL Nuclear Power Plant.)

Bupp, Irvin C., and Derian, Jean-Claude, *Light Water: How the Nuclear Dream Dissolved,* Basic Books, NY, 1978.

Caldicott, Helen, *Nuclear Madness, What You Can Do,* Autumn Press, Boston, 1978.

Cobb, J.C. (ed.), Surprising Findings about Plutonium Dangers to Man, Reported at the International Atomic Energy Meeting in San Francisco, CA, December 25, 1975.

Cohen, B.L., The Hazards in Plutonium Dispersal, Institute for Energy Analysis, March, 1975.

Cohen, B.L. The Disposal of Radioactive Wastes from Fission Reactors, *Scientific American,* 236-6:21, June, 1977.

Cohen, B.L., and response of Morgan, K.Z., and Rotblat, J. What Is The Misunderstanding All About? *Bulletin of Atomic Scientists,* 53-59, February, 1979.

Comey, D.D., The Legacy of Uranium Tailings, *Bulletin of Atomic Scientists,* 43-45, Sept. 1976.

Commoner, Barry, *Poverty of Power.* Alfred Knopf, NY, 1976.

Congressional Seminar on Low Level Radiation, Transcript of Proceedings, February 10, 1978.

Conrad, R.A., et al. A Twenty-Year Review of Medical Findings in a Marshallese Population Accidentally Exposed to Radioactive Fallout, Brookhaven National Laboratory, Associated Universities, Inc. USERDA.

Costa-Ribiero, C., et al. Radiobiological Aspects and Radiation Levels Associated with the Milling of Monzite Sands, *Health Physics,* 28:225, 1975.

Diamond, et al. The Relationship of Intrauterine Radiation to Subsequent Mortality and Development of Leukemia in Children, *Amer. J. Epid.,* 97:283, 1973.

DOE (United States Department of Energy) Report of Task Force for Review of Nuclear Waste Management, DOE/ER-0004, 1978.

EPA (United States Environmental Protection Agency), Magno, P.R., Reavey, T.C., Apidianakis, J.C. Iodine-129 in the Environment Around a Nuclear Fuel Reprocessing Plant (West Valley, NY), EPA Office of Radiation Programs, Field Operation Division, Washington, D.C., October, 1972.

EPA Environmental Analysis of the Uranium Fuel Cycle, EPA-520/4-76-017, 1976.

EPA, Terpilak, M.S. and Jorgensen, B.L. Environmental Radiation Effects in Nuclear Facilities in New York State, USEPA Radiation Data and Reports, 15-7:375, July, 1974.

EPA-ORP, Environmental Radiation Protection Requirements for Normal Operations of Activities in the Uranium Fuel Cycle, May, 1975.

EPA Estimate of the Cancer Risk Due to Nuclear Power Generation, USEPA, ORP/CSD-76-2, October, 1976.

EPA Radiation Protection Activities, 1976, EPA-520/4-77-005, August, 1977

EPA Natural Radioactivity Contamination Problems, EPA-520/4-77-015, February, 1978.

EPA Background Report: Considerations of Environmental Protection Criteria for Radioactive Waste, USEPA, February, 1978.

EPA Radiation Dose Estimates to Phosphate Industry Personnel, EPA-520/5-76-014.

EPA State of Geological Knowledge Regarding Potential Transport of High-Level Radioactive Waste from Deep Continental Repositories, EPA-520/4-78-004, 1978.

EPA Proceedings of a Public Forum on Environmental Protection Criteria for Radioactive Wastes, USEPA, ORP/CSD-78-2, May, 1978.

Faulkner, Peter (ed.). *The Silent Bomb: A Guide to the Nuclear Energy Controversy.* Vintage Books, NY, 1977.

Ford, D.F. Nuclear Power: Some Basic Economic Issues, Union of Concerned Scientists, 1975.

Ford, D.F. A History of Federal Nuclear Safety Assessments: From WASH-2740 Through the Reactor Safety Study, Union of Concerned Scientists, 1977.

Gilinsky, V. NRC Regulation of the Uranium Milling Industry: Problems and Prospects, May 2, 1978.

Gofman, J.W. and Tamplin, A.R. Epidemiologic Studies of Carcinogenesis by Ionizing Radiation, Proceedings of the 6th Berkeley Symposium on Mathematical Statistics and Probability, U.Cal.Press, 1971.

Gofman, J.W. The Fission-Product Equivalence Between Nuclear Reactors and Nuclear Weapons, Senate Congressional Record, Proceedings and Debates of the 92nd Congress, 1st Session, Vol. 117, July 8, 1971.

Gofman, John W. and Tamplin, Arthur R. *Poisoned Power.* Rodale Press, Emmaus, PA, 1971.

Gofman, J.W. The Cancer Hazard from Inhaled Plutonium, Committee for Nuclear Responsibility, CNR Report 1975-IR, Dublin, CA, May 14, 1975.

Gofman, J.W. The Cancer and Leukemia Consequences of Medical X-Rays, *Osteopathic Annals*, November, 1975.

Gofman, J.W. Estimated Production of Human Lung Cancers by Plutonium from Worldwide Weapons Test Fallout, Committee for Nuclear Responsibility, San Francisco, CA, July 10, 1975.

Gofman, J.W. The Plutonium Controversy, *J. American Medical Assoc.,* 236, July 19, 1976.

Gofman, J.W. Cancer Hazard from Low-Dose Radiation, CNR Report, 1977-9, NRC Docket No. RM 50-3, USNRC, 1977.

Gofman, J.W. Statement on lung cancer hazards of Pu-239 and Americium-241, Pamphlet, Women Strike for Peace, Washington, D.C., February 21, 1978.

Gorman, J. Wandering in the Nuclear Wasteland, *The Sciences,* 6-10, November, 1975.

Griem, M.L., et al. Analysis of the Morbidity and Mortality of Children Irradiated in Fetal Life, *Radiation Biology of the Fetal and Juvenile Mammal,* Proceedings of the 9th Annual Hanford Biology Symposium, ERDA, Richland, WA, May 5-8, 1969.

Guimond, R.J. Radiation and the Phosphate Industry—An Overview, 10th Midyear Topical Symposium of the Health Physics Society, Saratoga Springs, NY, October, 1976.

Gyorgy, Anna, et al. *No Nukes: Everyone's Guide to Nuclear Power.* South End Press, Boston, 1978.

Hart, J.C., Ritchie, R.H. and Varnadore, B.S. (eds.) Population Exposures, Proceedings of the 8th Midyear Topical Symposium of the Health Physics Society, Knoxville, TN, October 21-24, 1974.

HEW (Health, Education, and Welfare) BRH/NERHL 70-1, Estimate of Radiation Doses Received by Individuals Living in the Vicinity of a Nuclear Fuel Reprocessing Plant in 1968, HEW-PAS. EHS, May, 1970.

Honicker, Jeannine and Farm Legal. *Honicker vs. Hendrie: A Lawsuit to End Atomic Power.* Book Publishing Co., Summertown, TN, 1978.

Honicker, Jeannine and Farm Legal, An Older View Regarding Unalienable Rights, Honicker Staff Response to the NRC Denton Memorandum, January, 1979.

Ichikawa, S. and Nagata, M. Nuclear Power Plant Suspected to Increase Mutations, Laboratory of Genetics, Kyoto Univ., Kyoto, Japan, 1977.

Interagency Review Group, Report on Nuclear Waste Management, October, 1978.

Jordan, W.H. Memorandum to James R. Yore concerning errors in Table S-3, September 21, 1977, NRC.

Killough, G.G. and Till, J.E. Scenarios of C-14 Releases from the World Nuclear Power Industry from 1975 to 2020 and the Estimated Radiological Impact, *Nuclear Safety,* 19-5:602, 1978.

Kneale, G.W., Mancuso, T.F., Stewart, A.M., Cohort Study of the Cancer Risks from Radiation to Workers at Hanford (in press).

Kochupillai, N., Verma, I.C., Grewai, M.S., Ramalinga, Swami V. Down's Syndrome and Related Abnormalities in an Area of High Background Radiation in Coastal Kerala, *Nature,* 262, July 1, 1976.

Langham, W.H., Biological Implications of the Transuranium Elements for Man, *Health Physics,* 22:943-952, 1972.

Larson, R.T., and Oldham, R.D., Plutonium in Drinking Water: Effects of Chlorination on Its Maximum Permissible Concentration, *Science* 201, September 15, 1978.

Legator, M.S. and Hollander, A. Occupational Monitoring of Genetic Hazards, *Annals of the New York Academy of Sciences,* Vol. 269, December 31, 1975.

leVann, L.J. Congenital Abnormalities in Children Born in Alberta During 1961, *Can. Med. Assoc. J.,* 89:120, 1963.

Lisco, H., Finkel, M.P. and Brue, A.M. Carcinogenic Properties of Radioactive Fission Products and of Plutonium, *Radiology,* 49:361, 1947.

Little, J.B., et al. Distribution of Polonium in Pulmonary Tissues of Cigarette Smokers, *New England J. of Medicine,* 273:1343, 1965.

Lochstet, W.A. EIS Comments on Sweetwater Uranium Project and on Dalton Pass Uranium Mine, 1978.

Lorenz, E. Radioactivity and Lung Cancer, *J. National Cancer Inst.,* 5:1, 1944.

MacMahon, Pre-Natal X-Ray Exposure and Childhood Cancer, *J. National Cancer Inst.,* 28:1173, 1962.

Mancuso, T.F. Study of the Lifetime Health and Mortality Experience of Employees of ERDA Contractors, Dept. of Industrial Environmental Health Sciences, Grad. Sch. of Public Health, Univ. of Pittsburgh, PA, September 30, 1977.

Mancuso, T.F., Stewart, A. and Kneale, G. Radiation Exposures of Hanford Workers Dying from Various Causes, *Health Physics,* 33:369, November, 1977.

Martell, E.A. Iodine-131 Fallout from Underground Tests II, *Science* 148, June 25, 1965.

Martell, E.A. Actinides for the Environment and Their Uptake by Man, National Center for Atmospheric Research (NCAR), Boulder, CO, May, 1975.

Martell, E.A. Tobacco Radioactivity and Cancer in Smokers, *Amer. Scientist,* 63:404, July, 1975.

Matanoski, et al. The Current Mortality Rates of Radiobiologists and Other Physician Specialists: Specific Causes of Death, *Amer. J. Epidemiology,* 101:3, 1975.

Morgan, K.Z. Suggested Reduction of Permissible Exposure to Plutonium and Other Transuranium Elements, *Amer. Industrial Hygiene Assoc. J.,* August, 1974.

Morgan, K.Z. Reducing Medical Exposure to Ionizing Radiation, *Amer. Industrial Hygiene Assoc. J.,* May, 1975.

Morgan, K.Z. The Dilemma of Present Nuclear Power Programs, Presented at Hearings before the Energy Resources Conservation and Development Commission, Sacramento, CA, February 1, 1977.

Morgan, K.Z. The Linear Hypothesis of Radiation Damage Appears to be Non-Conservative in Many Cases, Proceeding IV International Congress of the International Radiation Protection Association, Paris, France, Vol. 2, April 26-27, 1977.

Morgan, K.Z. Cancer and Low-Level Ionizing Radiation, *Bulletin of Atomic Scientists,* 30-41, September, 1978.

Morgan, K.Z. Testimony presented before the Senate Government Affairs Committee, March 6, 1979.

Morgan, K.Z. Radiation Induced Cancer in Man, Subcommittee on Energy, Nuclear Proliferation and Federal Services, U.S. Senate, March 6, 1979.

Nader, Ralph and Abbots, John. *The Menace of Atomic Energy.* W.W. Norton & Co., Incl., NY, 1977.

Natarajan, N. and Bross, I.D.J. Preconception Radiation and Leukemia, *J. of Medicine,* 1973.

Neel, R.G., Soil-Relationships with Respect to the Uptake of Fission Products, U.S.A.E.C., UCLA 247, 1953.

NRC (United States Nuclear Regulatory Commission) GESMO, Generic Environmental Statement on the Use of Mixed Oxide Fuels in the LWR Fuel Cycle, NUREG-0002, USNRC, 1976.

NRC Kepford, C. Testimony and Appeals Briefs, in the Matter of Three-Mile Island Unit 2, NRC Docket 50-320.

NRC Three-Mile Island Nuclear Station Hearings before the NRC Licensing Board, 1977-78.

NRC, *Nuclear Safety* Staff, Radiological Quality of the Environment in the United States, 1977, *Nuclear Safety,* 19-5:617, 1978.

NRC Hearing on Health Effects of Ionizing Radiation, April 6, 1978.

NRC Response to the Honicker Petition: An Overview Regarding Exposure to Radiation (Denton Memorandum), November, 1978.

Nuclear Energy's Dilemma—Disposing of Hazardous Radioactive Waste Safely, Report to Congress by the Comptroller General of the United States, GAO-Government Accounting Office, September 9, 1977.

Olson, McKinley C. *Unacceptable Risk.* Bantam Books, NY, 1976.

O'Farrell, T.P. and Gilbert, R.O. Transport of Radioactive Materials by Jackrabbits on the Hanford Reservation, *Health Physics,* 29:9-15, July, 1975.

Oveharenro, E.P. An Experimental Evaluation of the Effects of Transuranium Elements on Reproductive Ability, *Health Physics,* 1972.

Patrick, C.H. Trends in Public Health in the Population Near Nuclear Facilities: A Critical Assessment, *Nuclear Safety,* 18(5):647, 1977.

Pohl, R.O. Health Effects of Radon-222 from Uranium Mining, *Search,* 7(5):345, August, 1976.

Polhemus, D.W. and Koch, R. Leukemia and Medical Radiation, *Pediatrics,* March, 1959.

Proceedings of a Congressional Seminar on Low Level Ionizing Radiation, Environmental Policy Center, November, 1976.

Proceedings of a Congressional Seminar on Federal Ionizing Radiation Standards, Environmental Policy Center, February 10, 1978.

Radford, E.P. Testimony before the House Subcommittee on Health and Environment, Hearings on Low Level Ionizing Radiation, February 9, 1978.

Richmond, C.R. and Thomas, R.L. Plutonium in Man and His Environment, *Nature,* 263-265, July 21, 1962.

Ryan Report, HR-95-1090, Nuclear Power Costs, House Committee on Government Operations, Report of the Subcommittee on Energy, Environment, and Natural Resources, April 26, 1978.

Sasaki, Yuichiro. *Hiroshima.* Gensuikin, 1977.

Scott, K.G., et al. Occupational X-Ray Exposure Associated with Increased Uptake of Rubidium by Cells, *Archives of Envir. Health,* 26:64, 1973.

Scott, R.L., and Gallaher, R.B. Events Resulting in Reactor Shutdown and Their Causes, *Nuclear Safety,* Vol. 20-1:92, 1979.

Segi, M. et al. Cancer Mortality in Japan, Dept. of Public Health, Tohoku University School of Medicine, Sendai, Japan.

Seltzer, R. and Sartwell, P.E. The Influence of Occupational Exposure to Radiation on the Mortality of Radiologists and Other Medical Specialists, *Amer. J. Epidemiology,* 81:2, 1965.

Sikov, M.R. and Mahlum, D.D. Plutonium in the Developing Animal, *Health Physics,* 22:707, 1972.

Smith, D.D.and Black, S.C. Actinide Concentrations in Tissues from Cattle Grazing Near the Rocky Flats Plant, National Environmental Research Center Report, NERC-LV-539-36, pp. 1-3.

Sternglass, E.J. The Role of Indirect Radiation Effects on Cell Membranes in Immune Response, and Evidence for Low Level Radiation Effects on the Human Embryo and Fetus, *Radiation Biology of the Fetal and Juvenile Mammal,* Proceedings of the 9th Annual Hanford Biology Symposium, ERDA, Richland, WA, May 5-8, 1969.

Sternglass, E.J. Environmental Radiation and Human Health, *Effects of Pollution on Health,* Vol. 6 in Proc. of 6th Berkeley Symposium on Mathematical Statistics and Probability, U. of Cal. Press, Berkeley, CA, 1972.

Sternglass, E.J. Radioactive Waste Discharges from the Shippingport Nuclear Power Station and Changes in Cancer Mortality, May 8, 1973.

Sternglass, E.J. Radioactivity (Chapter 15), *Environmental Chemistry,* J.O'M. Bockris (ed.), Plenum Press, NY, 1977.

Sternglass, E.J. Changes in Infant Mortality Patterns Following the Arrival of Fallout from the September 26, 1976 Chinese Nuclear Weapon Test, July 18, 1977.

Sternglass, E.J. Strontium-90 Levels in the Milk and Diet Near Connecticut Nuclear Power Plants, October 27, 1977.

Sternglass, E.J. Cancer Mortality Changes Around Nuclear Facilities in Connecticut, Testimony presented at a Congressional Seminar on Low-Level Radiation, Washington, D.C., February 10, 1978.

Stewart, A. A Survey of Childhood Malignancies, *Brit. Med. J.*, 1:1495, 1958.

Stewart, A. and Kneale, G.W. Radiation Dose Efects in Relation to Obstetric X-Rays and Childhood Cancers, *Lancet*, 1:1185, 1970.

Takahashi, C. and Ichikawa, S. "Variation of Spontaneous Mutation Frequency in Tradescantia Stamen Hairs Under Natural and Controlled Environmental Conditions," *Environmental and Experimental Botany*. 16:287-293, 1976.

Taylor, Vince, "Energy: The Easy Path", Pan Heuristics, Study for the U.S. Arms Control and Disarmament Agency, Jan. 1, 1979.

Telesky, L. Occupational Cancer of the Lung, *J. Industrial Hygeine and Toxicology*, 19:73.

Till, J.E., Hoffman, F.O., Dunning, D.E., Jr. "Assessment of TC-99 Releases to the Atmosphere - A Plea for Applied Research," ORNL/TM-6260, Contract No. W-7405-eng-26, June, 1978.

West Valley and the Nuclear Waste Dilemma, 12th Report by Committee on Government Operations Together with Additions and Dissenting Views, October 26, 1977, Union Calendar No. 95th Congress, 1st Session, Report No. 95-755.

UNSCEAR, Ionizing Radiation Levels and Effects, Vols. 1 and 2, Publication No. E, 72 IX, 17 U.N., NY, 1972.

Viadana, E. and Bross, I.D.J. Use of Medical History to Predict the Future Occurrence of Leukemia in Adults, *Prev. Med.*, 1974.

Wagoner, J.K., et al. Radiation as the Cause of Lung Cancer Among Uranium Miners, *New England J. of Medicine*, 273:181, 1965.

Wesley, J.P. Background Radiation as the Cause of Fetal Congenital Malformation, *Intern. J. Rad. Biol.*, 2:297, 1960.

Common Radionuclides

These are the principal isotopes which occur from the fission of uranium, their longevity, type of radiation they emit, and their abundance at the time they leave the reactor.

Atomic # (electrons)	Element	Abbr	Isotope Mass # (protons + neutrons)	Half-Life	Radiation Type	Activity (curies per metric ton of uranium)
1	Hydrogen (Tritium)	H	3	12 years	Beta	71
4	Beryllium	Be	10	2,700,000 years	Beta	-
6	Carbon	C	14	5,770 years	Beta	-
20	Calcium	Ca	41	100,000 years	Positron	-
26	Iron	Fe	59	45 days	Beta	-
27	Cobalt	Co	60	5 years	Beta, Gamma	-
28	Nickel	Ni	59	80,000 years	Positron	-
30	Zinc	Zn	65	145 days	Beta, Gamma	-
34	Selenium	Se	79	70,000 years	Beta	0.398
36	Krypton	Kr	85	10 years	Beta, Gamma	11,300
36	Krypton	Kr	90	33 sec.	Beta, Gamma	-
37	Rubidium	Rb	87	47,000,000,000 years	Beta	-
38	Strontium	Sr	89	53 days	Beta	718,000
38	Strontium	Sr	90	28 years	Beta	77,600
39	Yttrium	Y	90	64 hours	Beta, Gamma	80,700
39	Yttrium	Y	91	58 days	Beta	938,000
40	Zirconium	Zr	93	950,000 years	Beta	1.89
40	Zirconium	Zr	95	65 days	Beta	1,370,000
41	Niobium	Nb	93m	4 years	Gamma	0.145
41	Niobium	Nb	95	35 days	Beta, Gamma	1,380,000
42	Molybdenum	Mo	93	10,000 years	Positron	-
43	Technetium	Tc	99	210,000 years	Beta, Gamma	14.3
44	Ruthenium	Ru	103	40 days	Beta	1,220,000
44	Ruthenium	Ru	106	1 year	Beta	545,000
46	Palladium	Pd	107	7,000,000 years	Beta, Gamma	0.11
47	Silver	Ag	110m	249 days	Beta, Gamma	3,680
50	Tin	Sn	126	100,000 years	Beta	0.546
51	Antimony	Sb	125	2 years	Beta	8,700
52	Tellurium	Te	127m	105 days	Gamma, Beta	15,400
52	Tellurium	Te	129	67 min.	Beta	337,000
53	Iodine	I	129	17,200,000 years	Beta, Gamma	0.037
53	Iodine	I	131	8 days	Beta, Gamma	861,000
53	Iodine	I	134	52 min.	Beta, Gamma	-

54	Xenon	Xe	133	5 days	Beta, Gamma	
54	Xenon	Xe	137	4 min.	Beta, Gamma	
54	Xenon	Xe	138	14 minutes	Beta, Gamma	-
55	Cesium	Cs	134	2 years	Beta, Positron ive	246,000
55	Cesium	Cs	135	2,000,000 years	Beta, Gamma	0.286
55	Cesium	Cs	137	30 years	Beta, Gamma	108,000
58	Cerium	Ce	144	285 days	Beta	1,110,000
61	Promethium	Pm	147	2 years	Beta, Gamma	102,000
63	Europium	Eu	154	16 years	Beta, Positron, Gamma	6,990
63	Europium	Eu	155	2 years	Beta	7,480
82	Lead	Pb	210	21 years	Beta, Alpha	-
83	Bismuth	Bi	210 m	3,000,000 years	Alpha	-
84	Polonium	Po	210	138 days	Alpha	-
86	Radon	Rn	220	1 min.	Alpha, Positron ive	0.002
86	Radon	Rn	222	4 days	Alpha	-
88	Radium	Ra	224	4 days	Alpha	0.002
88	Radium	Ra	225	15 days	Beta	-
88	Radium	Ra	226	1,622 years	Alpha	-
90	Thorium	Th	228	2 years	Alpha	0.002
90	Thorium	Th	229	7,340 years	Alpha	-
90	Thorium	Th	230	80,000 years	Alpha	-
90	Thorium	Th	232	14 years	Alpha	-
90	Thorium	Th	234	24 days	Beta	0.314
91	Proactinium	Pa	226	2 min.	Alpha, Positron	-
92	Uranium	U	233	162,000 years	Alpha	-
92	Uranium	U	234	248,000 years	Alpha	0.751
92	Uranium	U	235	713,000,000 years	Alpha	0.017
92	Uranium	U	236	23,900,000 years	Alpha	0.268
92	Uranium	U	238	4,510,000,000 years	Alpha	0.314
93	Neptunium	Np	237	2,200,000 years	Alpha	0.333
94	Plutonium	Pu	236	285 years	Alpha	0.350
94	Plutonium	Pu	238	86 years	Alpha	2,720
94	Plutonium	Pu	239	24,390 years	Alpha	318
94	Plutonium	Pu	240	6,580 years	Alpha	477
94	Plutonium	Pu	241	13 years	Beta, Alpha	105,000
94	Plutonium	Pu	242	379,000 years	Alpha	1.38
94	Plutonium	Pu	243	5 years	Alpha	-
94	Plutonium	Pu	244	76,000,000 years	Alpha	-
95	Americium	Am	241	458 years	Alpha	85.9
95	Americium	Am	242	16 hours	Beta, Positron, Alpha, Gamma	63.400
95	Americium	Am	243	7,950 years	Alpha	18.1
96	Curium	Cm	242	163 days	Alpha	33.400
96	Curium	Cm	243	35 years	Alpha	3.71
96	Curium	Cm	244	18 years	Alpha	2.440
96	Curium	Cm	247	40,000,000 years	Alpha	-

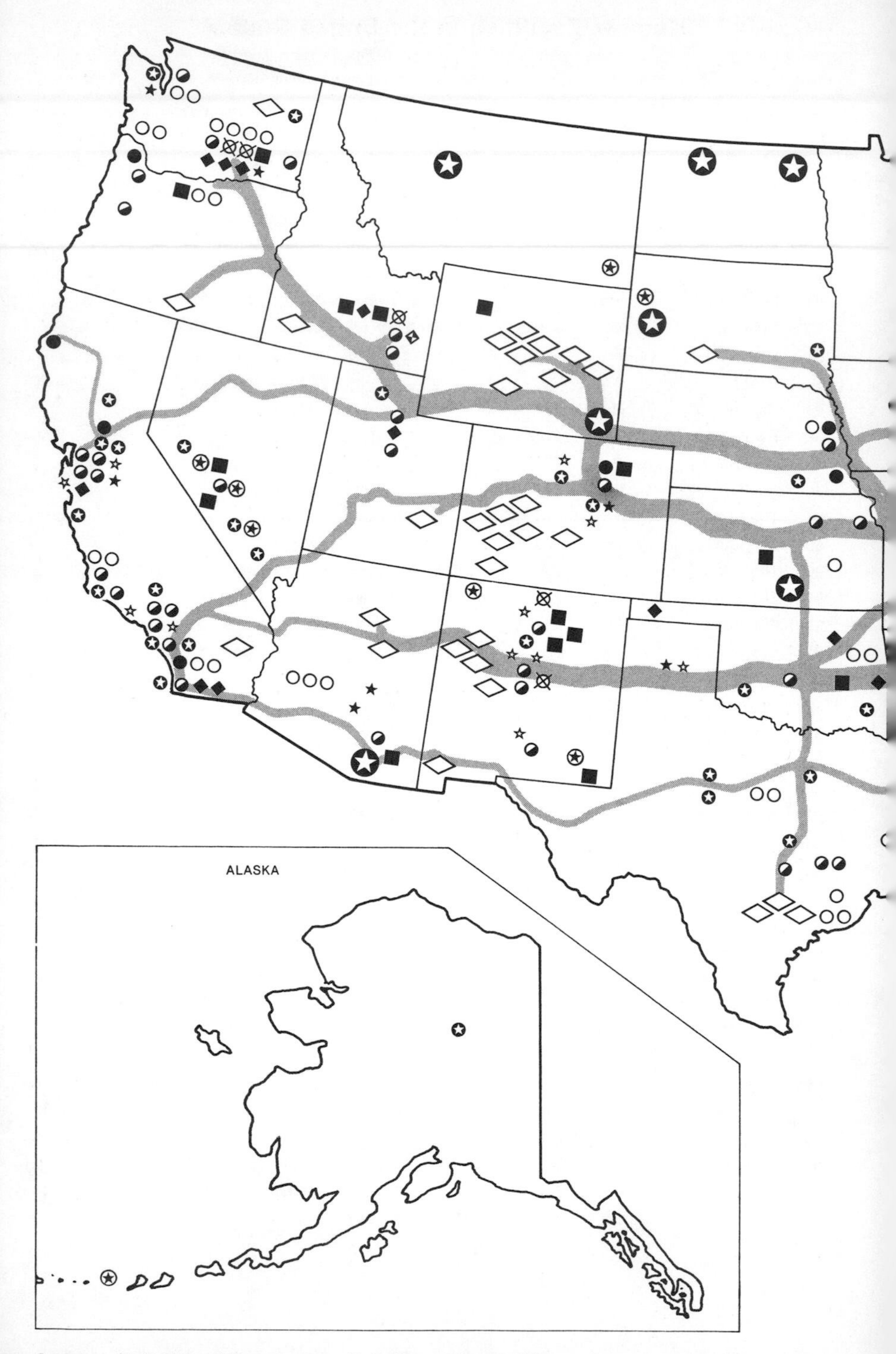

Sources: U. S. Dept. of Transportation, U. S. Dept. of Energy, War Resisters League, Another Mother For Peace, Hammond World Atlas.

Nuclear Facilities in the United States

MAJOR TRANSPORTATION ROUTES FOR NUCLEAR MATERIALS ARE INDICATED BY GREY LINES WITH WIDTH REPRESENTING APPROXIMATE VOLUME OF TRAFFIC.

JANUARY, 1979

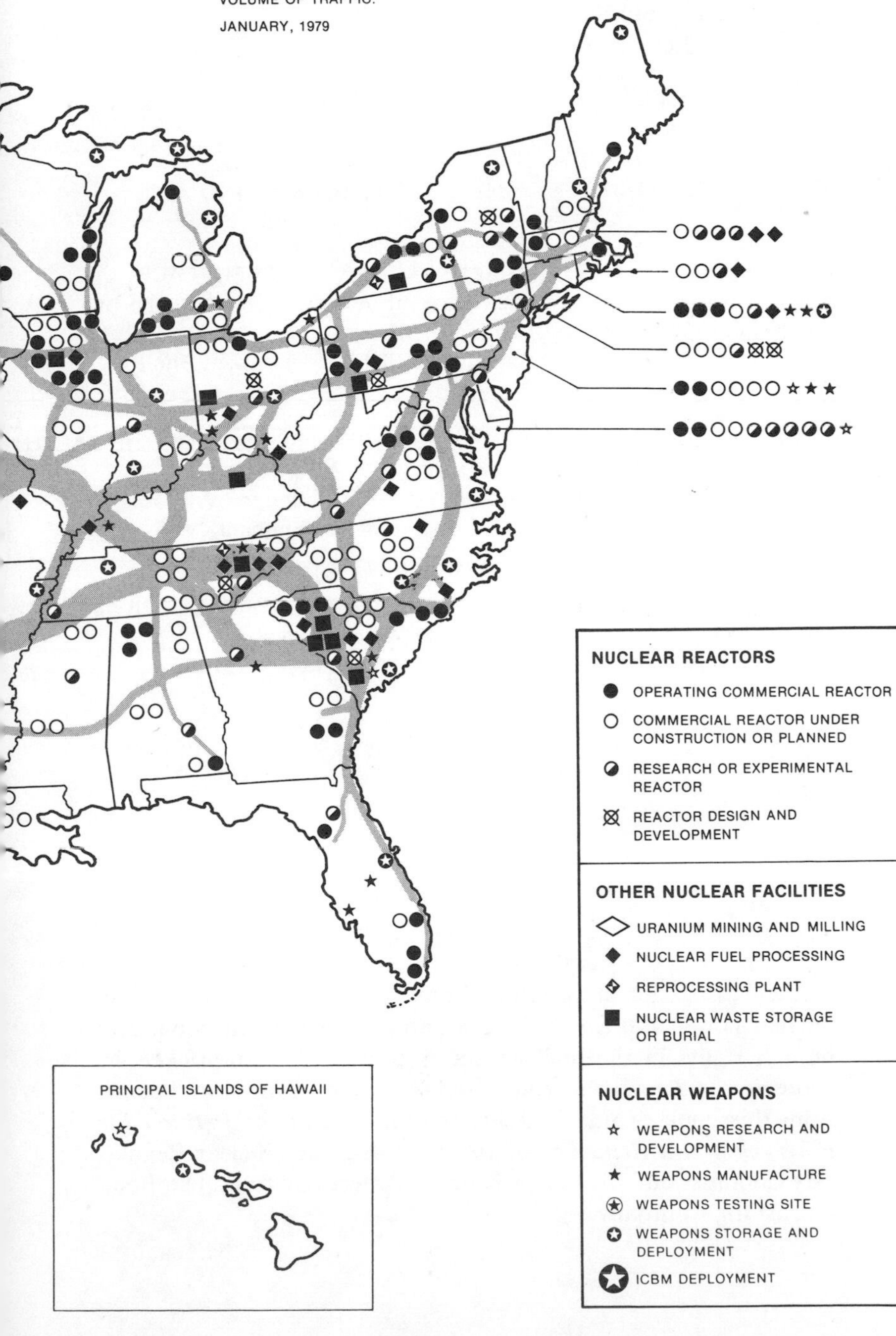

If you would like to help, or want more information, you can contact any of these groups below:

PLENTY
156 Drakes Lane
Summertown, TN 38483
(615) 964-3992

PLENTY
RR 3
Lanark, Ontario
Canada KOG 1KO

GREENPEACE
240 Fort Mason
San Francisco, CA 94123

GREENPEACE
2108 West 4th Ave.
Vancouver, B.C.
Canada V6K IN6

GREENPEACE
Club, Columbo St
London S.E.I.,
London, England

FRIENDS OF THE EARTH
620 C St. S.E.
Washington, D.C. 20003

FRIENDS OF THE EARTH
54-53 Queen St.
Ottawa, KIP 5C5
Canada

FRIENDS OF THE EARTH
9 Poland St.
London WIV 3DG
England

THE COUSTEAU SOCIETY INC.
777 3rd Ave.
New York, NY 10017

MOBILIZATION FOR SURVIVAL
3601 Locust Walk
Philadelphia, PA 19104

A full discussion of the scientific and medical facts surrounding nuclear power and a definitive argument for constitutional and human rights in the nuclear age is presented in ***Honicker vs. Hendrie: A Lawsuit to End Atomic Power.*** ***Shutdown!*** stems from this book, which contains the full text of the *Petition for Emergency and Remedial Action* filed with the Nuclear Regulatory Commission. Available from bookstores or from The Book Publishing Company.

You Can Make A Difference

Within our constitutional form of government, laws are made and changed by our political representatives. Since the Nuclear Regulatory Commission was established as an agency by the U.S. Congress, contacting your congressional representative is a particularly effective way to influence the outcome of the nuclear issue.

To send a copy of *Shutdown!* to the local, state, or federal official of your choice, send us their name, position, and address along with $4.00 for each book desired, and we will be happy to send them a copy. We will enclose your card for you, if you wish.

The Book Publishing Co.
156 Drakes Lane
Summertown, TN 38483

Project Director: Albert Bates, Farm Legal

Attorney: Joel Kachinsky, Farm Legal

Special thanks to: Dolph Honicker, Linda Honicker, John Gofman, Ernest Sternglass, Chauncey Kepford, Rosalie Bertell, Stephen Gaskin, The Farm Legal Staff, Sadeo Ichikawa, Kay Drey, Bob Alvarez, June Allen, Larry Bogart, Judith Johnsrud, William Lochstet, Irwin D.J. Bross, Thomas Mancuso, Leroy Ellis, Bob Pyle, Lewis Laska, Bill Garner, Karin Sheldon, Anthony Roisman, Lou Sirico, Robert Gary, Michael Bancroft, Ralph Nader, Franklin Gage, Harold Maier, Sam Lovejoy, Mr. and Mrs. Austin P. Malley, Helen Caldicott, Leo Goodman, Gertrude Dixon.

Published by The Book Publishing Company:

Publisher: Paul Mandelstein **Editor:** Matthew McClure **Contributing Editors:** Barbara Schaeffer, Dolph Honicker **Art:** Peter Hoyt, James Hartman, Mark Schlichting, Arthur Saarinen, Gregory Lowry **Composing:** Jane Ayers, Dana Gaskin, Carolyn James, Marcia McGee **Layout:** Tortesa Livick, David Long, June Setesak, James Egan **Darkroom:** Brian Hansen, Jenny Banks, Vance Glavis, Daniel Luna **Lithography:** Jeffrey Clark, Thomas Malamed-Durocher **Printing and Production:** Robert Seidenspinner, John Seward, Richard Martin, Albert Livick, Steve McGee, Keith Martin, Michael Tassone **Photographs:** Rusty Honicker **Distribution:** Bruce Moore

The research for this project was funded by PLENTY, a non-profit, tax-deductible, charitable organization. Contributions can be sent to PLENTY, 156 Drakes Lane, Summertown, Tennessee 38483